HOME

The **Beginning** Guide to

Microsoft Word 2007

Exam 77-601 Study Guide

another
Computer Mama Guide

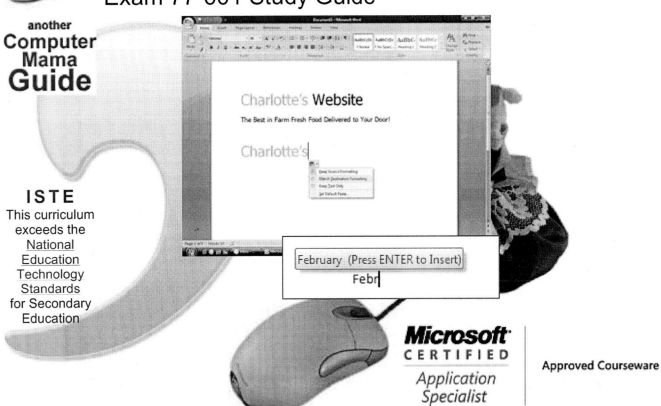

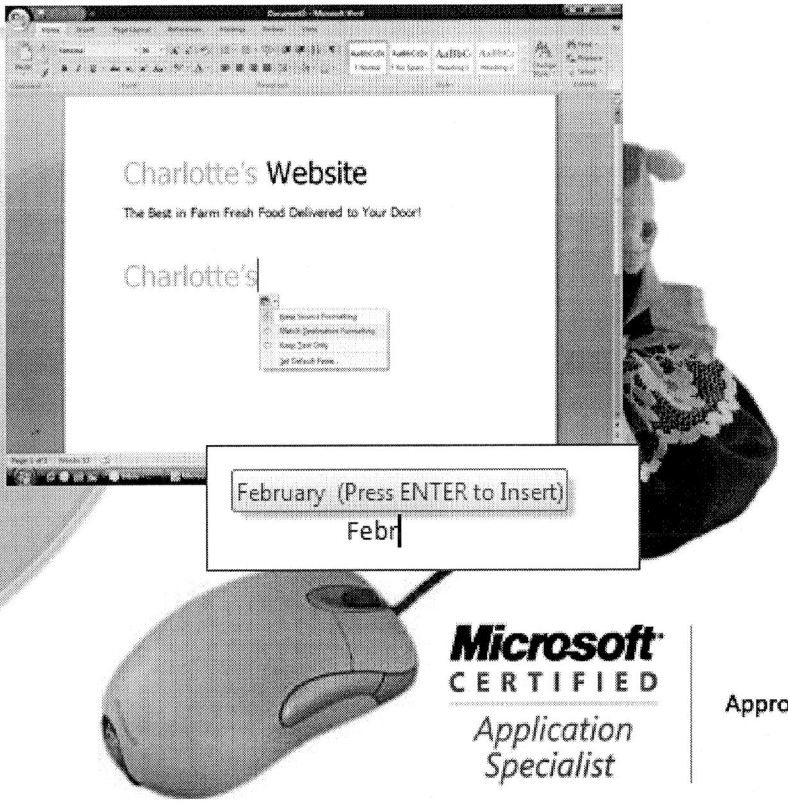

ISTE
This curriculum exceeds the National Education Technology Standards for Secondary Education

Microsoft CERTIFIED Application Specialist

Approved Courseware

The Comma Project

Beginning Guide to Microsoft Word 2007

another
Computer Mama Guide

9090 Chilson Road
Brighton, MI 48116
ISBN: 978-0-9818778-9-1

HOME

Comma Products

About this course

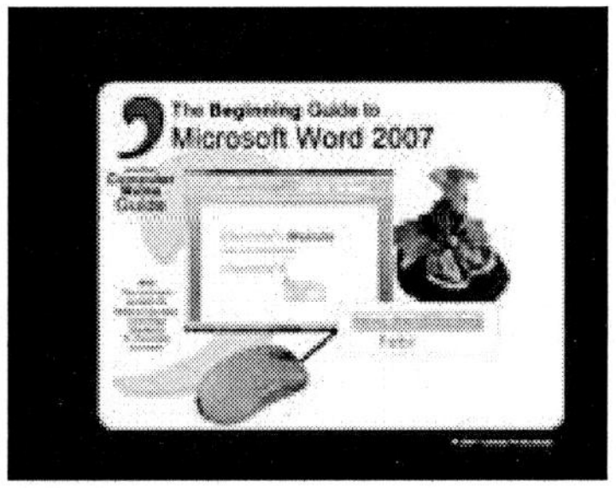

Comma Project, LLC.

Books Available in this Series:
Beginning Guide to Microsoft® Word 2007

Intermediate Guide to Microsoft® Word 2007

Advanced Guide to Microsoft® Word 2007

The Beginning Guide to Microsoft Word 2007
Exam 77-601: Using Microsoft® Office Word 2007

Description: *The Beginning Guide to Microsoft Word 2007* demonstrates how to edit and format text, insert pictures from ClipArt : Edit Undo, Redo, Copy, Paste, drag and drop editing, move text, use the mini-toolbar an auto text options.

Who will benefit from this course? Students who want to learn editing and layout will enjoy creating professional business documents with Microsoft Word 2007. Advanced students who are familiar with previous versions of Microsoft Word will quickly find the information they need to upgrade their skills.

Audience Description: Who should take this course?
The certification training is available to adult learners online in partnership with major colleges and universities. The audience for this course includes:

- Office workers, managers, entrepreneurs, teachers, and military personnel who want to start using advanced skills immediately
- Job training and professional development (WIA), as well as retired and unemployed people looking to expand their job possibilities

Course Prerequisites: Students who enroll in Microsoft Certified Application Specialist (MCAS) program should have basic computer skills including how to turn on the computer, how to use an Internet browser and how to select commands from a menu.

Online Course Requirements

Microsoft Certified Application Specialist (MCAS) certification training

You will need the following Microsoft products already installed on your computer in order to take this course online: Windows Vista Business edition, or Windows XP, For the Microsoft Office 2007 certification course, you MUST have the following software: Word 2007, Excel 2007, Access 2007; Outlook 2007and PowerPoint 2007. Microsoft Office 2007 is NOT the same as Microsoft Office 97-2003.

Hardware requirements for Vista Business:

- IBM-compatible (PC) computer running
- Processor: 1 GHz 32-bit (x86) or 64-bit (x64)
- RAM: 1 GB of system memory **(needs more)**
- Hard Drive: 40 GB with at least 15 GB of available space
- Video: Support for DirectX 9 graphics with WDDM Driver
- 128 MB of graphics memory (needs more)
- Pixel Shader 2.0 in hardware
- 32 bits per pixel
- DVD-ROM drive
- Audio Output

Adobe Acrobat Reader (free version) and a Flash Player.

An online course requires reliable, effect Internet access. If your internet service provider uses only dial-up, a minimum of 56K connection rate is recommended; however, high speed access (Cable or DSL) is preferred. This course cannot be taken with a Macintosh computer

Microsoft Certified Application Specialist

Microsoft CERTIFIED Application Specialist

Microsoft CERTIFIED Application Professional

For more information:
Microsoft Business Certification

What is the Microsoft Business Certification Program?

The Microsoft Business Certification Program enables candidates to show that they have something exceptional to offer – proven expertise in Microsoft Office programs. The two certification tracks allow candidates to choose how they want to exhibit their skills, either through validating skills within a specific Microsoft product or taking their knowledge to the next level and combining Microsoft programs to show that they can apply multiple skill sets to complete more complex office tasks. Recognized by businesses and schools around the world, over 3 million certifications have been obtained in over 100 different countries. The Microsoft Business Certification Program is the only Microsoft-approved certification program of its kind.

What is the Microsoft Certified Application Specialist Certification?

The **Microsoft Certified Application Specialist** Certification exams focus on validating specific skill sets within each of the Microsoft® Office system programs. The candidate can choose which exam(s) they want to take according to which skills they want to validate. The available Application Specialist exams include:

Using Microsoft ®Windows Vista™
Using Microsoft® Office Word 2007
Using Microsoft® Office Excel® 2007
Using Microsoft® Office PowerPoint® 2007
Using Microsoft® Office Access 2007
Using Microsoft® Office Outlook® 2007

Please Note: Comma Project, LLC. is independent from Microsoft Corporation, and not affiliated with Microsoft in any manner. While the Complete Computer Guides may be used in assisting individuals to prepare for a Microsoft Business Certification exam, Microsoft, its designated program administrator, and Comma Project, LLC. do not warrant that use of this Complete Computer Guides will ensure passing a Microsoft Business Certification exam.

Exam 77-601: Using Microsoft® Office Word 2007

Microsoft Certified Application Specialist (MCAS) reference topics

Description: The Microsoft Certified Application Specialist program is the only comprehensive, performance-based certification program approved by Microsoft to validate business computer skills using Microsoft Windows Vista® and Microsoft Office® 2007 productivity software: Excel, Word, Power Point, Access, and Outlook.

***The Beginning Guide to Microsoft Word** 2007* demonstrates how to Cut, copy, and paste, Insert pictures from ClipArt, Undo, Redo,Drag and drop editing, Move text, Insert ClipArt, and Resize pictures, Insert date and time, Insert picture from file, Format font and font size, File Save As a template.

The Intermediate Guide to Microsoft Word 2007 demonstrates how to Use Mail Merge: Create main document, Create data source, Sort records to be merged, Merge the document and data, Create and format tables, Add borders and shading, Merge Cells, Design a Web Page, Create Hyperlinks, Use Design Gallery Live.

The Advanced Guide to Microsoft Word 2007 demonstrates how to Create an on-line form with drop down lists and default text values, Create and edit styles, Create a Table of Contents, Use Headers and Footers, Create Section Breaks, Use the Document Map.

Microsoft Certified Application Specialist (MCAS) objectives for Word 2007

Study Guides
Beginning Word
Intermediate Word
Advanced Word

Microsoft Word Study Guide

Microsoft Certified Application Specialist (MCAS): Microsoft Word 2007 Exam 77-601 Guide

Table of Contents

Beginning Guide to Microsoft® Word 2007

Table of Contents, continued

Beginning Guide to Microsoft® Word 2007

Memo to self

Exploring Commands

The very first command in this Guide is on the **Home Ribbon.** The upper left corner of the **Home** toolbar is where the basic commands are found.

Question: Did you notice that all Microsoft Office 2007 programs have the same Home commands? Make your own **Menu Map** and use it to document where the tools are.

Getting Started

Welcome

Click Here to Get Started

Objectives

Lesson objectives: Learn effective methods for using online certification training. At the end of this lesson you will be able to:

Welcome

Intro to Computers presents a practical, hands-on approach to computers. The lessons are based on what you see on the screen, what you can do with the options, and what works on the job. The goal is to enable you to use Microsoft Windows and Office 2007 effectively, even creatively.

Use this *Guide* as part of your professional development plan to prepare for the new Microsoft Business Certification (Microsoft Certified Application Specialist MCAS) or as a reference book to solve problems as they come up.

This introduction provides information on:
- Navigation
- Practice
- Sample Documents
- Assessments

Log into the course

This online course requires a User Name and Password. You probably received an email with your username and password when you enrolled.

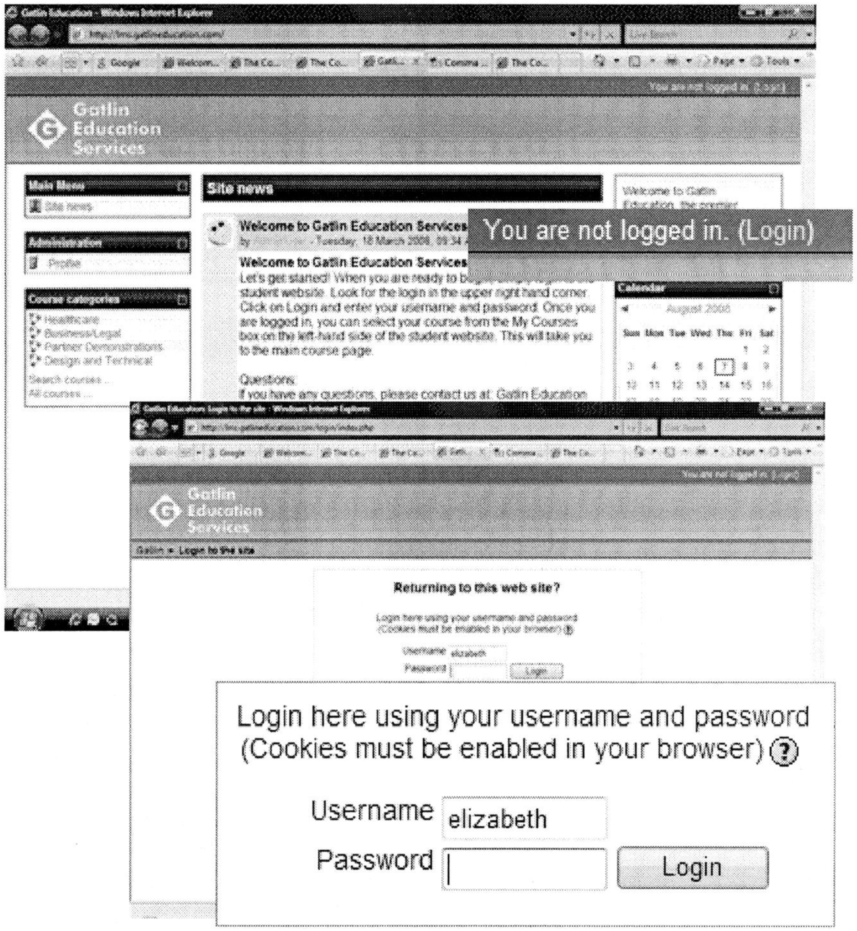

Try This: Login

Go to the website indicated in your Welcome email. When you are at the website, click on the **(Login)** link. You will be prompted for your Username and Password.

What If This Doesn't Work?

First, look at the keyboard and make sure the Caps Lock is off (no light.) Passwords need both upper and lower case letters.

Second, check the spelling. Your user name may not be exactly the same as your E-mail address.

Third, you can click on the Live Chat and get immediate assistance.

Intro to Computers ->Topic Outline

The Topic Outline

The **Topic Outline** lists your lessons, practice and quizzes. Each Level has six lessons that you should read.

For each Level, there are 5 or 6 links to the material you should review. There are additional reference materials and practice tests you can use if you wish. Copies of these reference materials are reprinted in back of the hard copy books.

This *Computer Guide* is a hard copy print out of the online course. Certainly, you can use the *Guides* to practice offline, especially if you are entering data into a spreadsheet or designing a flier.

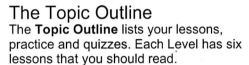

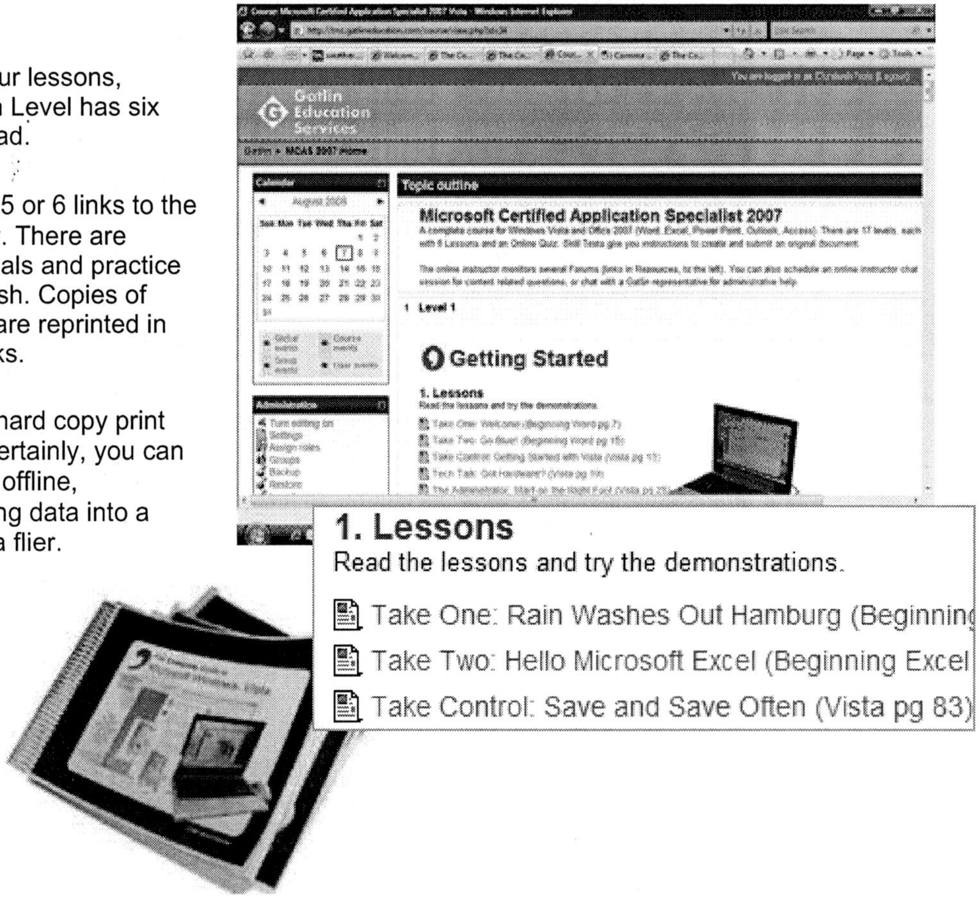

Intro to Computers ->Topic Outline ->Lesson

Lesson Links

When you click on a hyperlink to read a lesson, a new window will open. At the top of the window you should see how many **pages** are in this lesson. You can click on the page numbers to go through the lesson. You can also use the white arrows on the right side of the screen to go **next** and **back** .

What Do You See? When you are done with a lesson, you can close the browser window. Go to the upper right corner of the lesson window and click on the **X** to **Exit**.

The **Topic Outline** should be the there, the window left open.

Intro to Computers ->Topic Outline -> Downloads

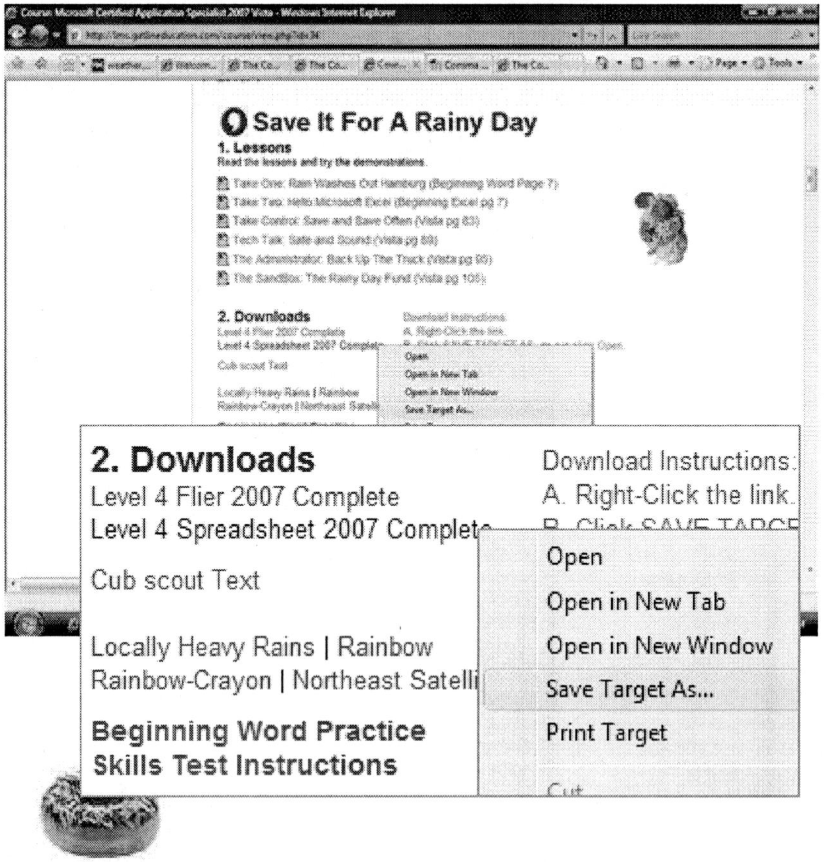

Download the Samples

You can often learn more by "taking a thing apart" to see how it was made. This online course includes sample documents, spreadsheets and pictures that you can download if you wish.

Try This: Download a File
Login the course online.
Go to the **Topic Outline**.
Scroll down to the **Downloads**.
Right Click the Skill Test Instructions.
Left Click **Save Target As**.
Go to your **Documents** folder.
Click on **Save**.

What Do You See? Right Click gives you a list of options. In the option list you should see **Save Target As**. The words might not match if you have a different browser than the one shown on this page.

Memo to Self: Come back to this page again when you need to: after you go through the sections on Saving Files later in the *Guide*.

Take a Quiz Online

After you review the materials online or with the *Guides*, you can log into the course online and take a **Quiz**. This is an open book quiz. You are allowed to look up the answers in your notes, online, or in the computer *Guides*.

Review the Quiz Buttons
Submit: This button posts your answer for the current question.

Save without submitting: This button saves your answers. You can leave the quiz and finish it later.

Save without submitting

Submit page: This button sends your answers to all questions on the page.

Submit all and finish: Use this button to finish the quiz and post your results online.

Submit all and finish

Intro to Computers -> Topic Outline -> Quiz

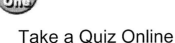

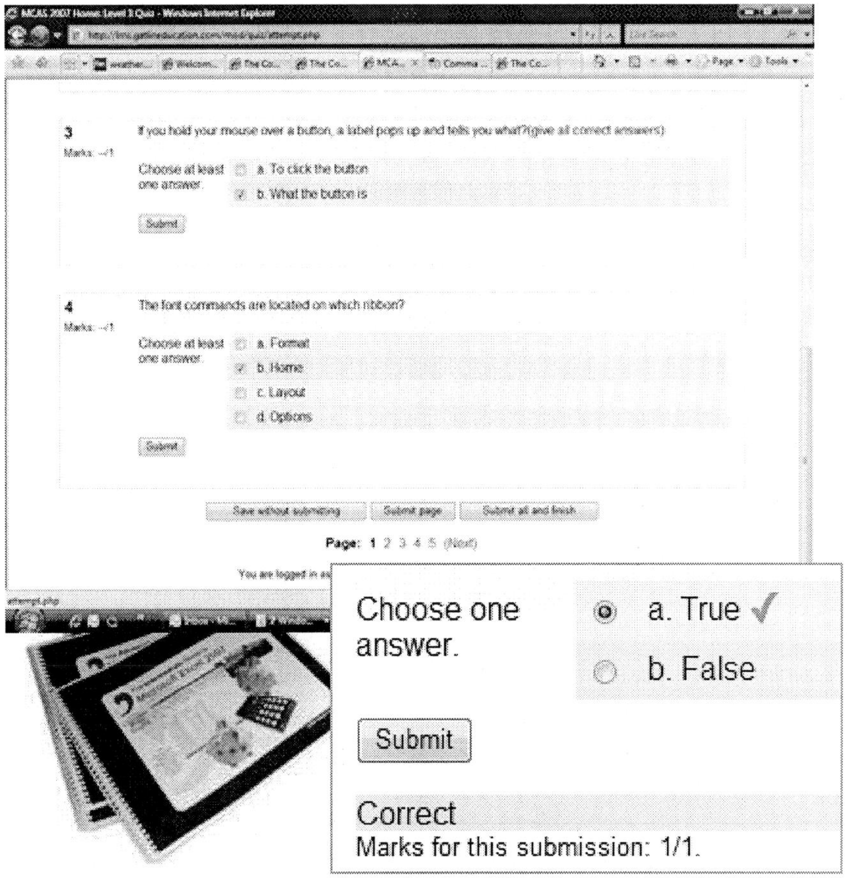

Submit Your Work

Many online courses ask you to **upload** a document or a spreadsheet. Here are the steps.

Try This: Upload a file
Go to the **Topic Outline**.
Click on **Submit...**

What Do You See? The instructions should be repeated on the screen.

Click on **Browse** to select the file you want to upload.Navigate to your file.

Click **Open**. The path and file name appear in the upload box.

Click **Upload this file** to submit the file to your instructor.

Memo to Self: Come back to this page again when you need to: after you go through the sections on Finding Files later in the *Guide*.

Intro to Computers -> Topic Outline -> Upload a file

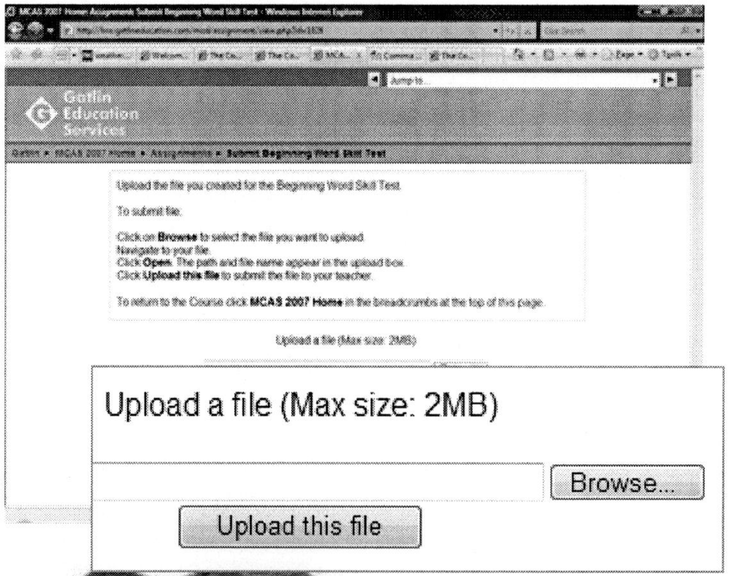

Intro to Computers -> Activities

Use the Forums

In an online class, a **Forum** is similar to raising your hand and asking a question. When you post a question to a Forum anyone can reply with a suggestion or comment. Some of the answers are very creative and useful.

Your instructor may also post an explanation or offer additional links.

Live Chats

Many instructors keep Office Hours. Chat allows you to type questions online and get an answer immediately from your instructor when your instructor is in the office.

Don't Explain and Don't Complain:
Please keep your posts professional and on topic!

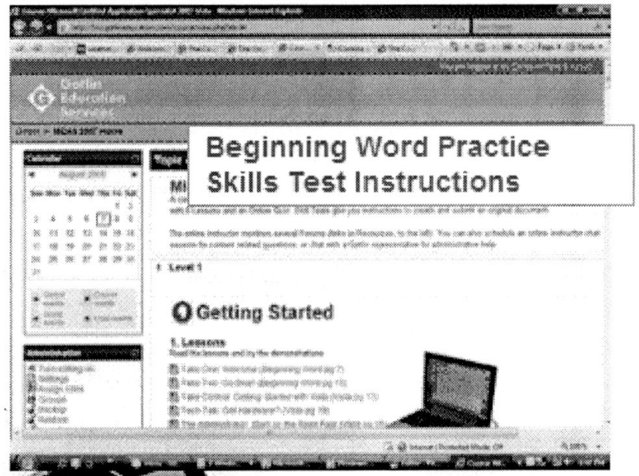

Practice

This *Guide* offers additional reference materials and practice certification tests. You can use the multiple choice quizzes and skill tests to practice if you wish. When you are ready, please log into the course and do the assessments online.

The Microsoft certification test are timed: you have to perform the process steps very quickly and efficiently in order to pass.
That takes practice!

More practice

If you have a question about a document or file you are working on you are always welcome to email a copy of your work to your instructor as an attachment.

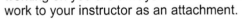

Take One

Microsoft CERTIFIED Application Specialist

Microsoft CERTIFIED Application Professional

What is the Microsoft Business Certification Program?

The Microsoft Business Certification Program enables candidates to show that they have something exceptional to offer – proven expertise in Microsoft Office programs. The two certification tracks allow candidates to choose how they want to exhibit their skills, either through validating skills within a specific Microsoft product or taking their knowledge to the next level and combining Microsoft programs to show that they can apply multiple skill sets to complete more complex office tasks. Recognized by businesses and schools around the world, over 3 million certifications have been obtained in over 100 different countries. The Microsoft Business Certification Program is the only Microsoft-approved certification program of its kind.

What is the Microsoft Certified Application Specialist Certification?

The **Microsoft Certified Application Specialist** Certification exams focus on validating specific skill sets within each of the Microsoft® Office system programs. The candidate can choose which exam(s) they want to take according to which skills they want to validate. The available Application Specialist exams include:

Using Microsoft ®Windows Vista™
Using Microsoft® Office Word 2007
Using Microsoft® Office Excel® 2007
Using Microsoft® Office PowerPoint® 2007
Using Microsoft® Office Access 2007
Using Microsoft® Office Outlook® 2007

Please Note: Comma Project, LLC. is independent from Microsoft Corporation, and not affiliated with Microsoft in any manner. While the Complete Computer Guides may be used in assisting individuals to prepare for a Microsoft Business Certification exam, Microsoft, its designated program administrator, and Comma Project, LLC. do not warrant that use of these Complete Computer Guides will ensure passing a Microsoft Business Certification exam.

Microsoft Business Certification

The *Complete Computer Guide* prepares you to pass the **Microsoft Certified Application Specialist** exam. This credential recognizes the business skills needed to get the most out of Microsoft Office **2007** .

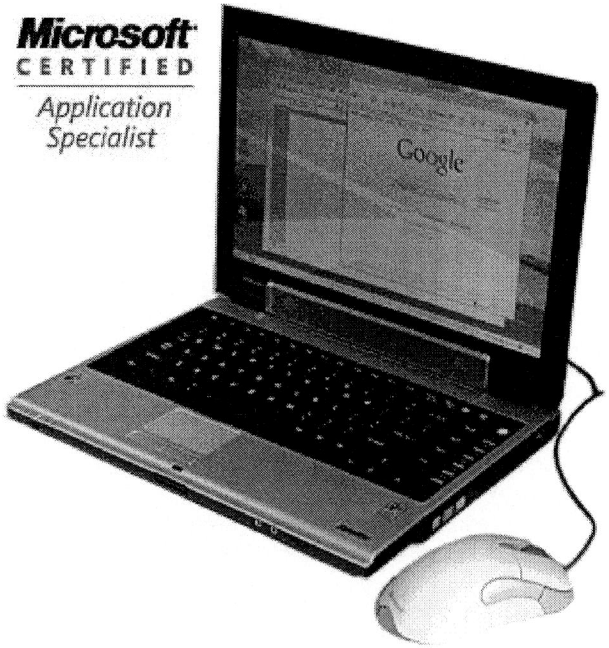

Microsoft Certification Tests are available through authorized testing centers. They are not included as part of the certification training program in the same way that taking the Bar Exam is not included with getting a degree in Law from college.

More Information Online.

Certiport provides the official Microsoft certification tests. Here is their address: www.certiport.com

You can download the Microsoft certification topics and study guides.

Learn Now, Earn Now

► Microsoft Office products are used in over 97% of U.S. businesses.

► Sixty-nine percent (69%) of office jobs require some knowledge and expertise in the use of Microsoft Office products.

► Microsoft has issued over 3 million MOS 2003 (Microsoft Office 2003) and MCAS 2007 (Microsoft Office 2007) certificates worldwide.

► Students who pass the MCAS certification exams **earn more (about 12%)** than employees who are not certified.

► In addition, **82%** of the students who get certified report **getting a raise after completing their certification training.**

► Many employers consider certification in determining who to hire – MCAS certification can be the difference in whether or not you get a better job.

Especially for women:

► More women (56.7%) than men (44.1 %) use computers at work.

► Approximately 75% of Microsoft Office certifications are granted to women.

► The Computer Mama™ courses are designed **especially for women** – the Comma Method of training recognizes learning differences for women and incorporates them into the courseware design. As a result, women have much greater success in these courses, when compared to traditionally designed courses. Men learned equally as well in Comma Method courses.

Self-Assessment

Skill Level-Beginning	Mastered	Needs Work	Required for my job
Create a new document			
Select, copy and paste text			
Format text			
Format columns			
Format borders and shading			
Spell and Grammar Check			
Insert a picture from ClipArt or a file			

Beginning Word is recommended if you selected “needs work” on three or more skills

Skill Level-Intermediate	Mastered	Needs Work	Required for my job
Create a watermark			
Use Headers and Footers			
Create a template			
Create a table			
Add, delete and modify rows, column and cells			
Create a Mail Merge			
Insert a bookmark or a hyperlink			
Create an on-line (Web) page			

Intermediate Word is recommended if you selected “needs work” on three or more skills

Skill Level-Advanced	Mastered	Needs Work	Required for my job
Format text with Styles			
Navigate with the Document Map			
Insert captions, footnotes or endnotes			
Create a Table of Contents			
Track Changes			
Create an on-line form			
Create a Master Document			

Advanced Word is recommended if you selected “needs work” on three or more skills.

Getting Started

Go Blue!

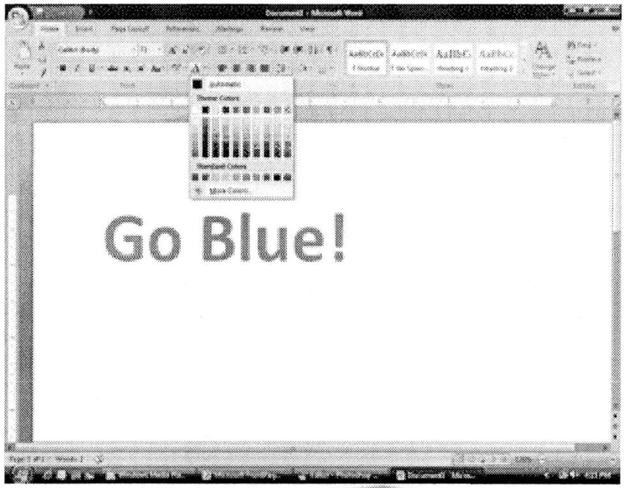

Click Here to Get Started

Beginning Word

Lesson objectives: To be able to create documents with rich text. At the end of the lesson you will be able to:

Open Microsoft Word and identify the Home Ribbon page 2

Demonstrate how to enter and edit text page 4

Understand how to select text page 6

Use the Home Ribbon to format text page 7

Use the Home Ribbon to format a paragraph page 11

Identify the options for formatting the Line Spacing page 12

© 2009 Comma Productions

Hello Microsoft Word!

Say, “Hello” to Microsoft Word. The *Complete Computer Guide* introduces Microsoft Word at the Newbie level, for folks who are new to computers. If you are farther along, that’s OK. You can still pick up some tricks and nomenclature.

Start -> All Programs -> Microsoft Office -> Microsoft Office Word 2007

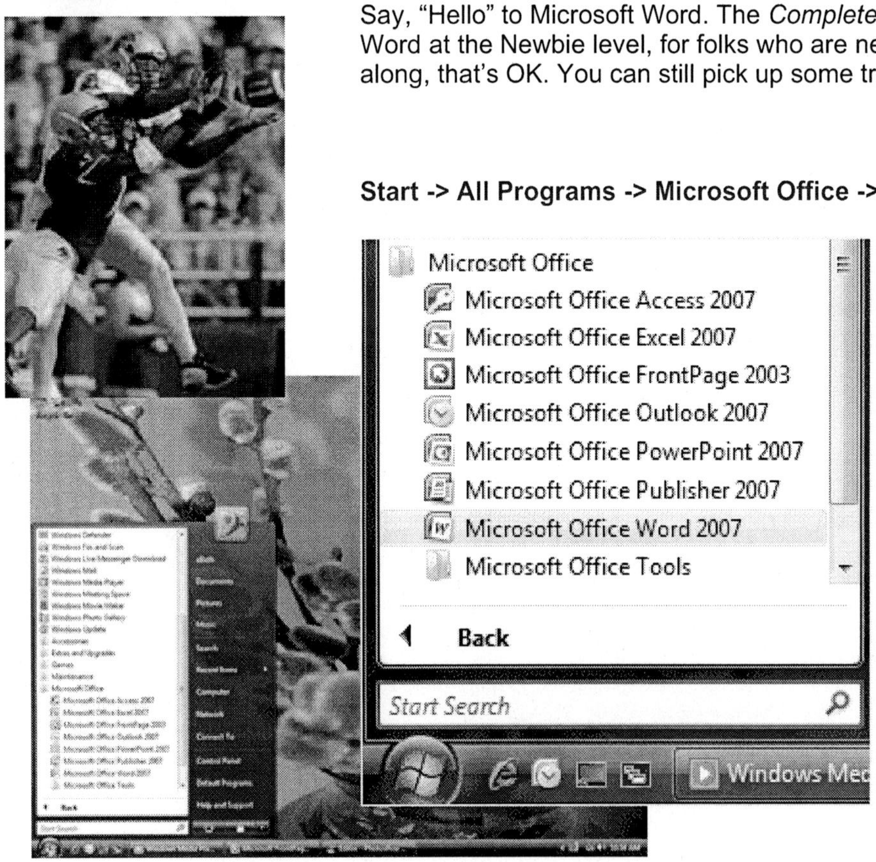

Look at the label in **BOLD** above the picture of the Start menu. This label is a method for writing out the path that your mouse travels. You will see this **map** at the top of every graphic.

Try it: Start Microsoft Word
Go to **Start**
Click on **All Programs**
Go to **Microsoft Office**
Click on **Microsoft Office Word 2007**

Start -> All Programs -> Microsoft Office -> Microsoft Word 2007

What do you see from the top of the screen? Is there a **Title Bar** that says Microsoft Word? Yes.

Is there a **Home Ribbon** with the Clipboard, Font and Paragraph Groups? Yes.

If your screen looks similar to the example on this page, then you are ready to get started.

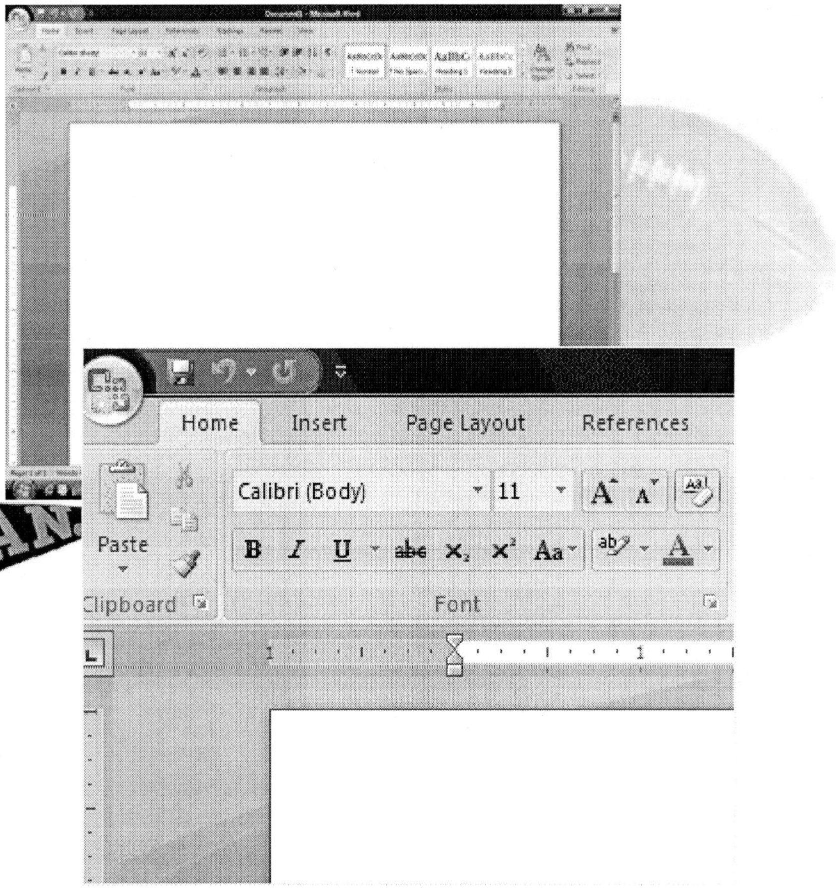

Take Two

Microsoft Word 2007

Getting Started

Microsoft Word, Microsoft Works, and Word Perfect are all word processors. The original purpose of a word processor was to work with text. Many of the steps for entering text mimic an old typewriter: word processors competed with typewriters in the early days. This lesson is basic: can you type in text? Can you select the text and format it?

Try it: Enter Type
Type the words: Go Blue.
Press the **Enter** key on the keyboard.

Using Enter is like hitting the return key on an old typewriter. It gives you a new line.

Move your mouse around the page. The mouse looks like an I-beam. Click your mouse at the end of the word "Blue." Do you see a flashing vertical line? That's the **cursor**, insertion point. Whatever you type next will start from that point.

Delete Type
Click your cursor at the end of the word "Blue."
Use the **backspace** key on your keyboard.
The word is removed one letter at a time.

Text Editing

Microsoft Word includes a 20,000 word dictionary. The word processor compares your typing to the words listed in the dictionary. If Word cannot find a match, you will see a red, wavy line.

Try it: Correct a spelling mistake
Type in the word: Miichigan
Place your cursor between the two "i's" and backspace or delete the extra letter.

Not bad for a start. But we live in the age of color, graphics, and CNN. What can you do to make this more exciting?

Microsoft Word

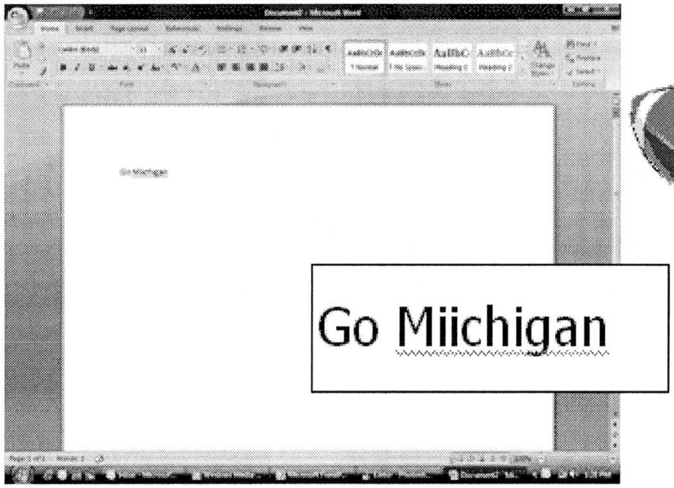

Microsoft Word

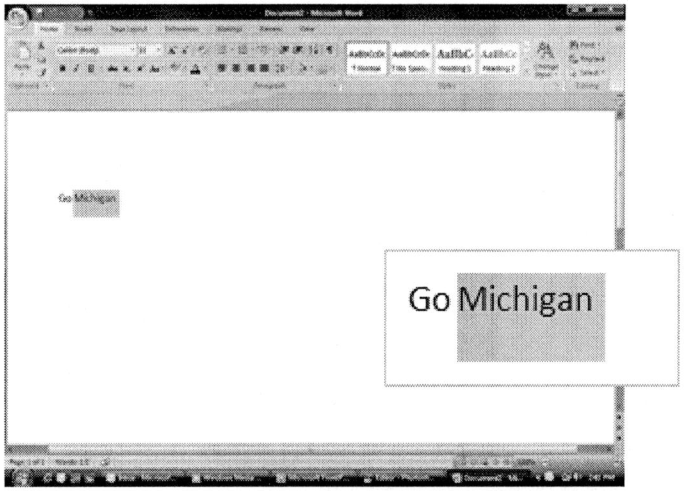

Select the Text

Nothing happens on a computer until you select it. Double click on the word, “Michigan.” to select it. You should see a blue box around the text when it is selected.

Single click just places your cursor.

Double click on a word to select it.

Triple click to select a paragraph.

Old tricks: To select a line of type, place your cursor between the edge of the paper and the first letter of type for the sentences you want to highlight. The cursor changes from an I-beam to a white arrow. Click and hold your left mouse button to drag and select the type you want.

Format the Text

Change the Size

Use the **Font** group on the **Home** Ribbon to make the word "Michigan" big, bold and blue. The default Font size is 11 pts. Look for the small down arrow by the number 11 and select 28 from the list.

Make It Bold

Go to the **Home** Ribbon
Click on "B" for **Bold** on the Font Toolbar.

Make It Blue

Now, make it blue by going to the button on the far right: it's a letter "A" with a color bar underneath. See the little arrow to the right ? Click on it to bring up the color palette.

The **Live Preview** in Office 2007 lets you see your changes as you sweep the color palette with your mouse.

WYSIWYG: What you see is what you get.

Home -> Font

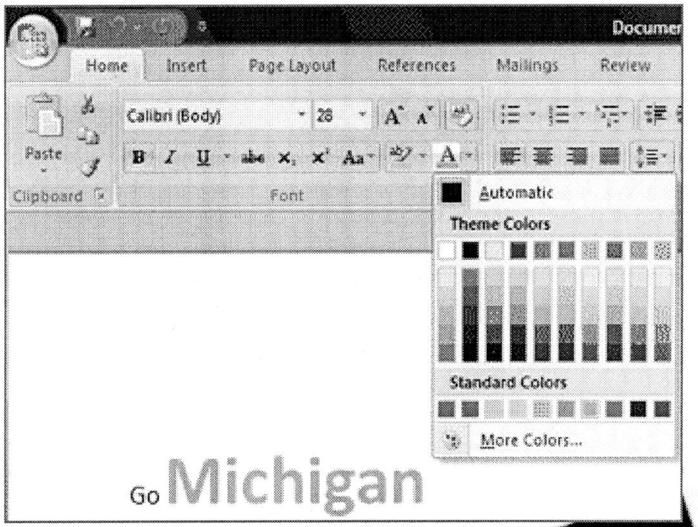

Microsoft Word 2007 Exam 77-601 Topic: 2. Formatting Content
2.1. Format text and paragraphs
2.1.3. Format characters: Change font size

Home -> Font

Format the Fonts

A **FONT** is a type face. Formatting the text with a bold or decorative Font distinguishes your document.

For example, a business letter should have a simple font that is easy to read. A wedding invitation has a different audience and a different purpose. Your invitation might have a handwritten font that looks as elegant as calligraphy.

Try This: Change the Font
Double click the word Michigan. This will highlight the text and **select** it.

Go to **Home ->Font**.
The default font in Microsoft Word 2007 is called Calibri. Click on the down arrow on the left and select a different font from the list.

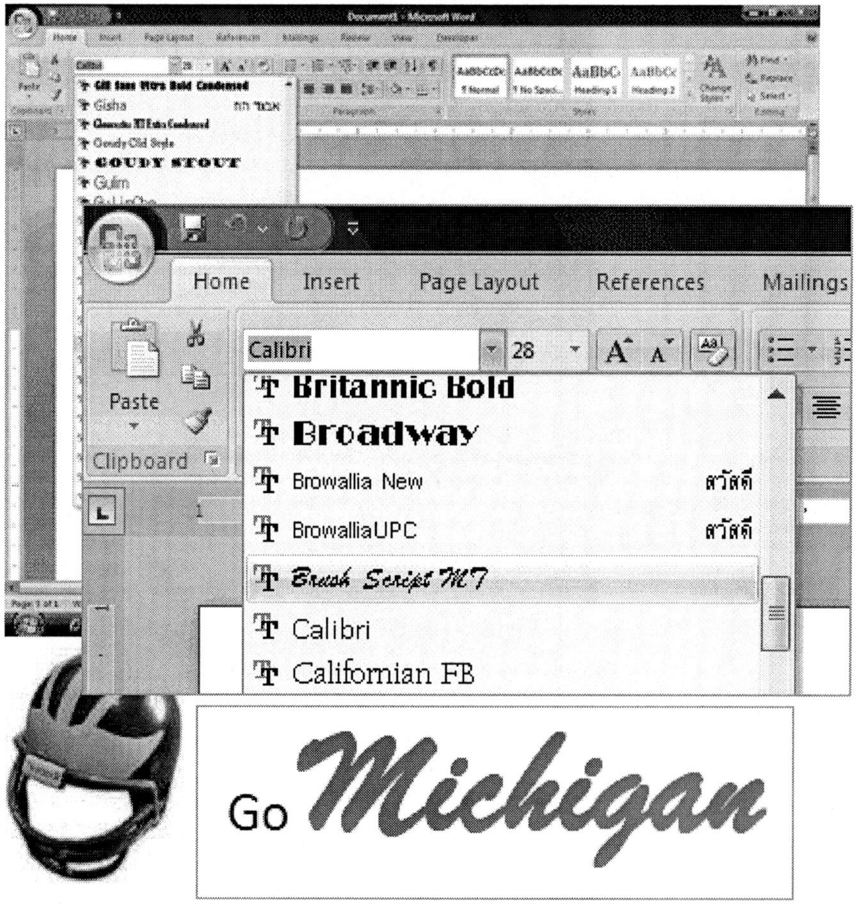

Microsoft Word 2007 Exam 77-601 Topic: 2. Formatting Content
2.1. Format text and paragraphs
2.1.3. Format characters: Change fonts

Home -> Font ->Change Case

More Format Options

You can change whether the text is Upper or Lower case with the Home Ribbon, too.

Try This, Too: Change the Font Case
Select the text: Michigan.
Go to **Home->Font ->Change Case.**
The **Change Case** button looks like a Capital "A" and and small "a."

What Do You See? The options include:

Sentence Case: Make the first letter in the sentence big, upper case.

Lowercase: All small, lower letters.

Uppercase: All big, upper letters.

Capitalize: Make the first letter of each word big, upper case.

Toggle: Change Upper to Lower and vice versa. You would use this to change the text if you left the Caps Lock on the keyboard.

Microsoft Word 2007 Exam 77-601 Topic: 2. Formatting Content
2.1. Format text and paragraphs
2.1.3. Format characters: Change font case

Clear the Formatting

Say you wanted to remove all of the formatting that you added to the text. This option can be useful when you copy and paste information from a web page.

Try This: Clear the Formatting
Sample the text: Michigan
Go to **Home ->Font ->Clear Formatting.** The Clear Formatting button is in the upper right corner of the Font group.

What Do You See? The text will return to the default Font (Calibri) and size (11 pt.)

Home -> Font ->Clear Formatting

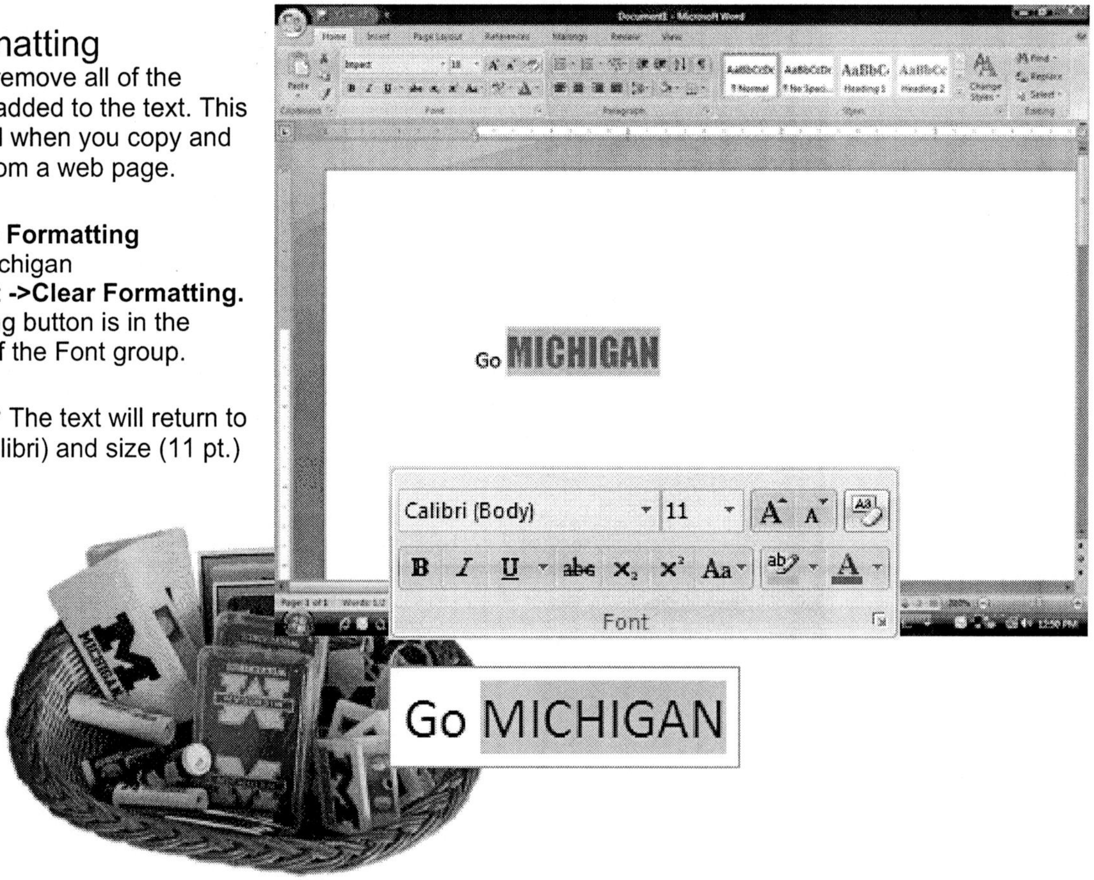

Home -> Paragraph

Format the Paragraph

Change the Alignment

Triple click the word Michigan.
Use the **Paragraph** group on the **Home** Ribbon to align the words “Go Michigan” to the **center** of the first line.

Left, Center and **Right** are kind of obvious alignments. **Justified** looks like the columns in a printed newspaper. Justified text is distributed evenly within the column.

The default alignment for letters, books and reports is Left with a "ragged right" edge.

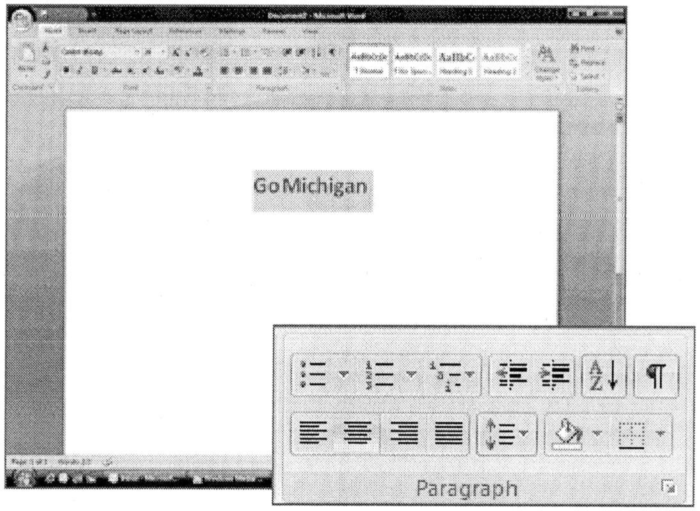

Microsoft Word 2007 Exam 77-601 Topic: 2. Formatting Content
2.1. Format text and paragraphs
2.1.4. Format paragraphs: Change alignment

Home -> Paragraph

Format the Paragraph

The amount of space between the lines of text is technically called **leading** in the print business. For several hundred years each letter of type was made from wood or metal and placed into a key frame. The rows of type were separated by strips of thin, lead bar: the leading. In Microsoft Word the leading is called **Line Spacing.**

Before You Begin
Place your cursor before the M of Michigan
Click ENTER on the keyboard to force a new line. Now you have two lines of type.

Try This: Increase the Line Spacing
Select both lines of type by dragging the mouse over the text.
Go to the **Home** Ribbon
Click **Line Spacing** and try different sizes.

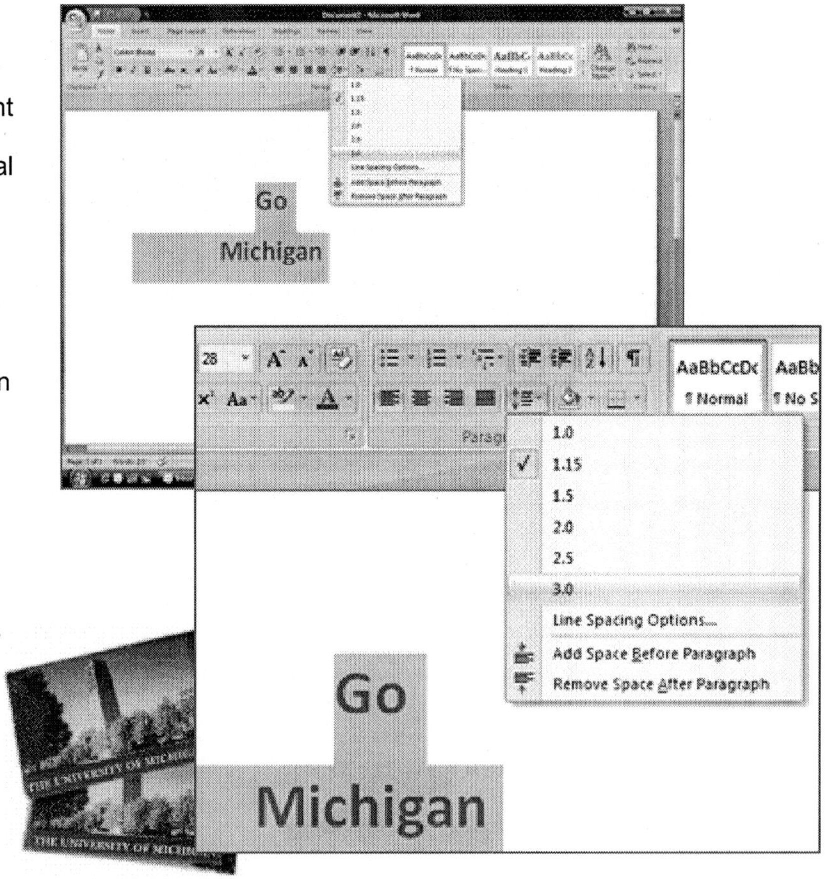

Microsoft Word 2007 Exam 77-601 Topic: 2. Formatting Content
2.1. Format text and paragraphs
2.1.4. Format paragraphs: Change line spacing

HOME
DONE

Rich Text Formatting

And that's all there is to it.

What you have been doing is called **Rich Text**. It's good for getting attention in your letters or e-mail. "Mama, send money!"

It's also a fun way to get to know text editing in Microsoft Word.

Well, you done good. You get the cookie. <grin>

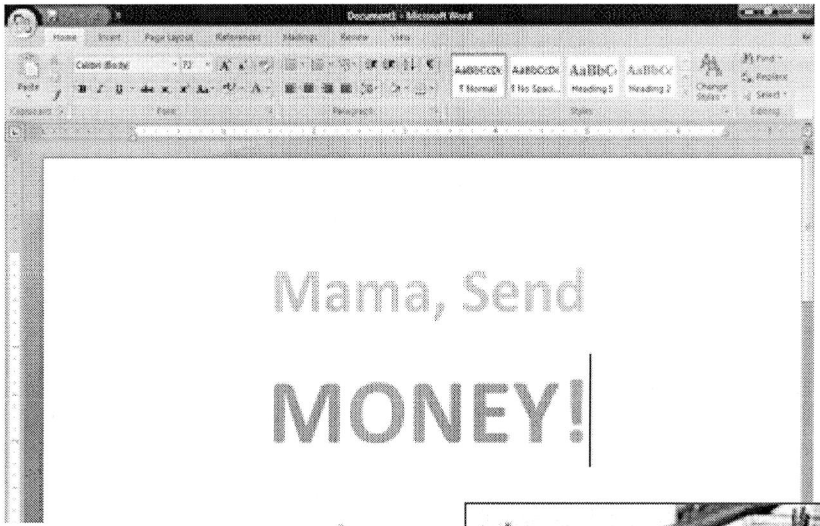

Test Yourself

1. In the screenshot below, what is the arrow is pointing to?

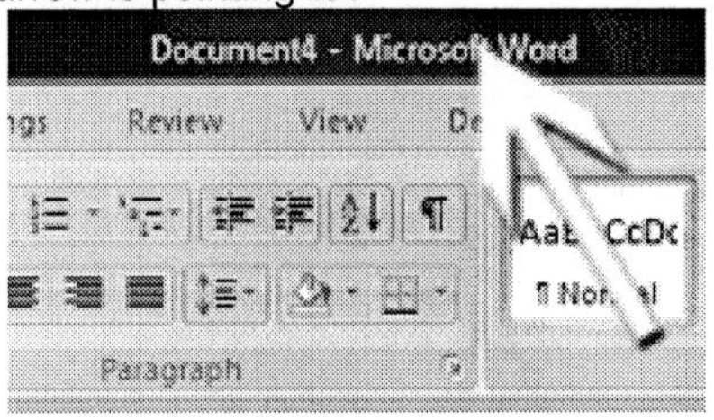

a. Title Bar
b. Heading
c. Window frame
Tip: Beginning Word: page 27

2. In the screenshot below, what ribbon is selected?

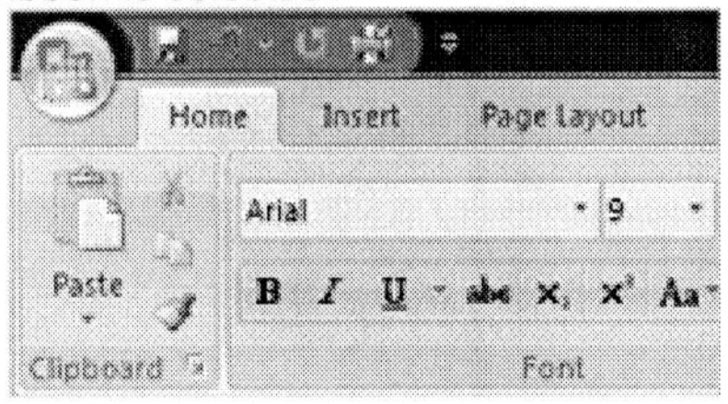

a. Insert
b. Normal
c. Home
d. View
Tip Beginning Word: page 27

3. Which of the following are Font Case options?
(select all correct answers)
a. Upper Case
b. Lower Case
c. Capitlize Each Word
d. tOGGLE cASE
Tip: Beginning Word: page 33

4. In the screen shot below, what does the circled button do?

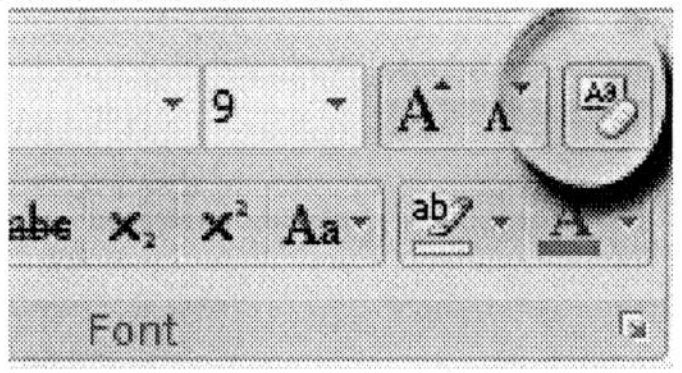

a. Erase text
b. Clear formatting from text
Tip: Beginning Word: page 34

5. In the screenshot below, what does the circled button do?
(select all correct answers)

a. Adjust line spacing
b. Makes the text double spaced
c. Adjusts font size
Tip: Beginning Word, page 36

Samples

There are no samples for this lesson.

Assessment

There are no **Practice Activities** or **Skill Test** for this lesson. You can continue to the next lesson if you wish.

Mice and Men

Horses and Zebras

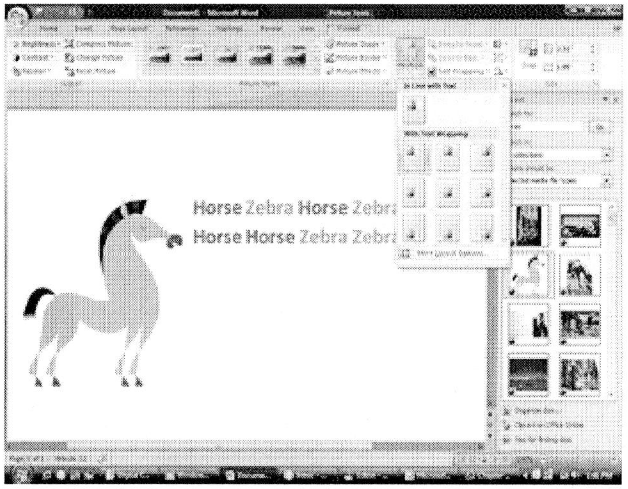

Click Here to Get Started
Sample Files

Beginning Word

Lesson Objectives: This lesson demonstrates how to use copy and paste with text as well as pictures. At the end of this lesson you will be able to:

Identify the Home Ribbon and locate the Groups page 2

Demonstrate how to enter, select and format text page 5

Use the Home Ribbon to Copy and Paste text page 6

Demonstrate how to move text Drag and Drop editing page 7

Use the Insert Ribbon to add ClipArt page 8

Use the ClipArt search options page 9

Select a picture and use Picture Tools page 11

Demonstrate how to resize pictures page 12

Use the Home Ribbon to Copy and Paste pictures page 13

Demonstrate how to use your mouse to move pictures page 14

Start Microsoft Word

The purpose of this lesson is to learn how to use the basic edit commands: **Copy**, **Paste** and the Computer Mama's favorite: **Undo**. If you were familiar with Microsoft Word in a previous version, this is a good place to get reacquainted with the new toolbars. So, **Start** the **Program** Microsoft Word.

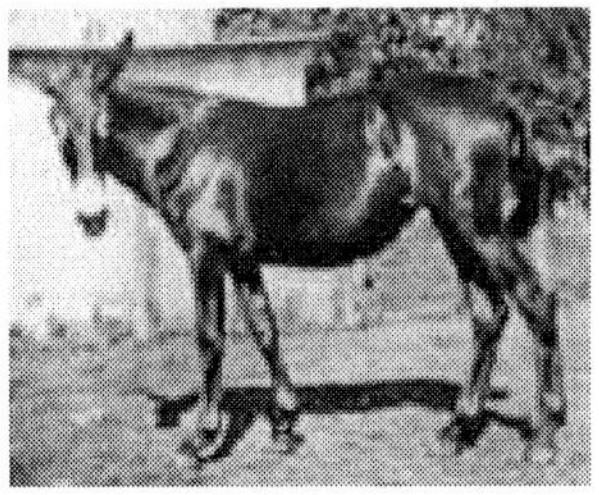

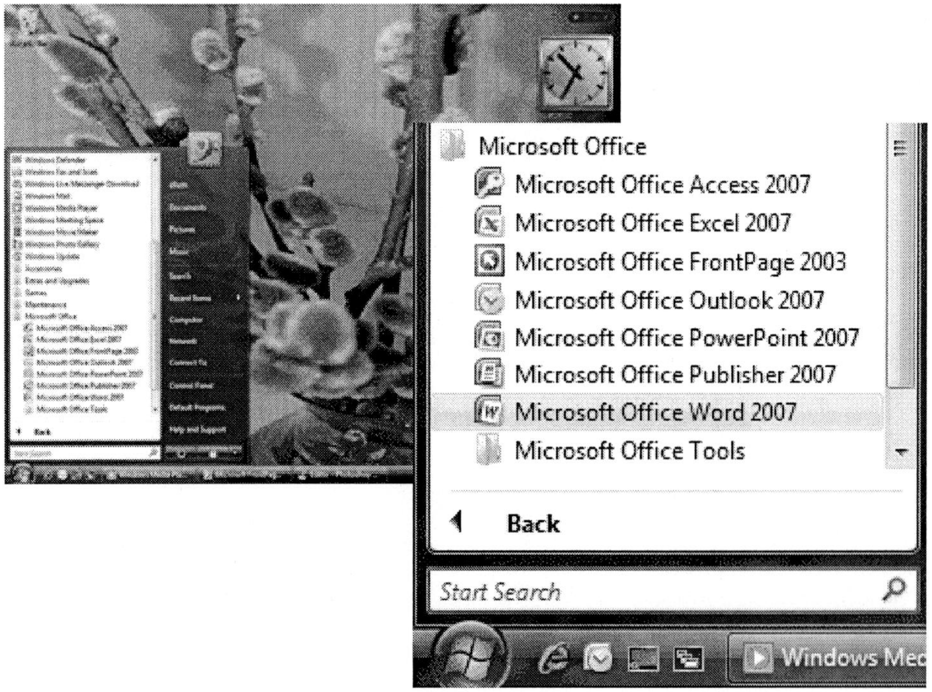

Start -> All Programs ->Microsoft Word

Where Are You?

What do you see from the top of the screen? Is there a **Title Bar** that says Microsoft Word? Yes.

Is there a **Home** Ribbon with the Clipboard, Font and Paragraph Groups? Yes.

If your screen looks similar to the example on this page, then you are ready to get started.

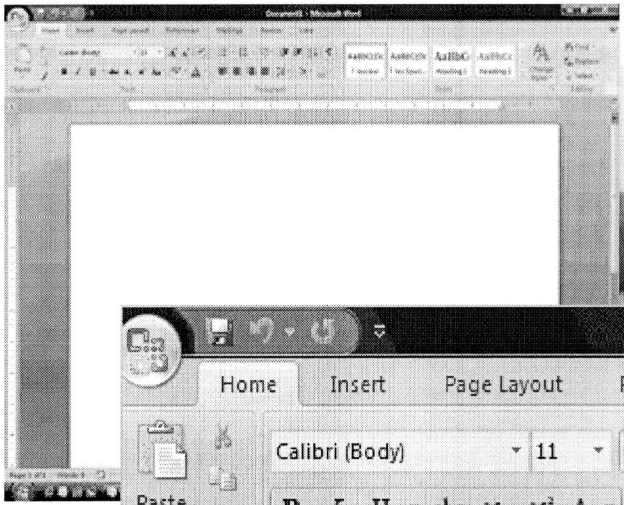

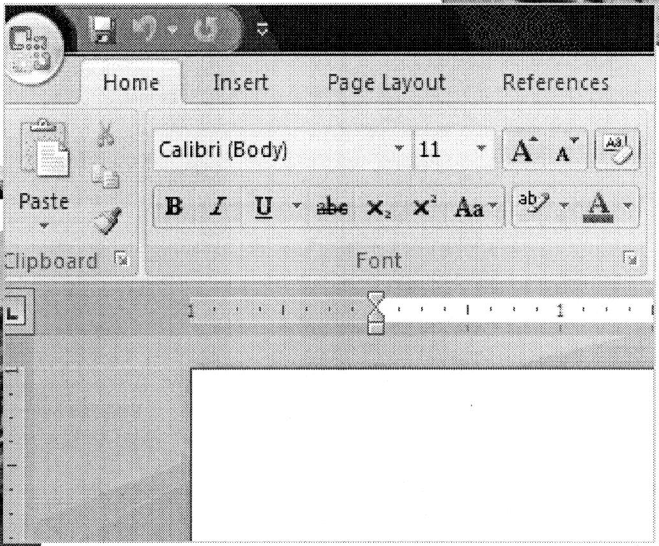

Where Are You Going?

Microsoft Office 2007 was redesigned based on user experiences. The design teams determined the tools that people used the most and placed them "up front." The menus were replaced with Ribbons. Ribbons are **Tabs** of functions that work together.

For example, all of the options for making a Mail Merge have been gathered on the **Mailings** tab.

Some tool bars are activated when you need them. When you click on a picture, the **Picture Tools** are displayed. When you click back on text, the Picture Tools are turned off.

Each **Ribbon** may have many **Groups**, and **Commands**. In the example on this page, the **Font** group includes Bold, Italic, Underline, etc.

Microsoft Word 2007 -> Home

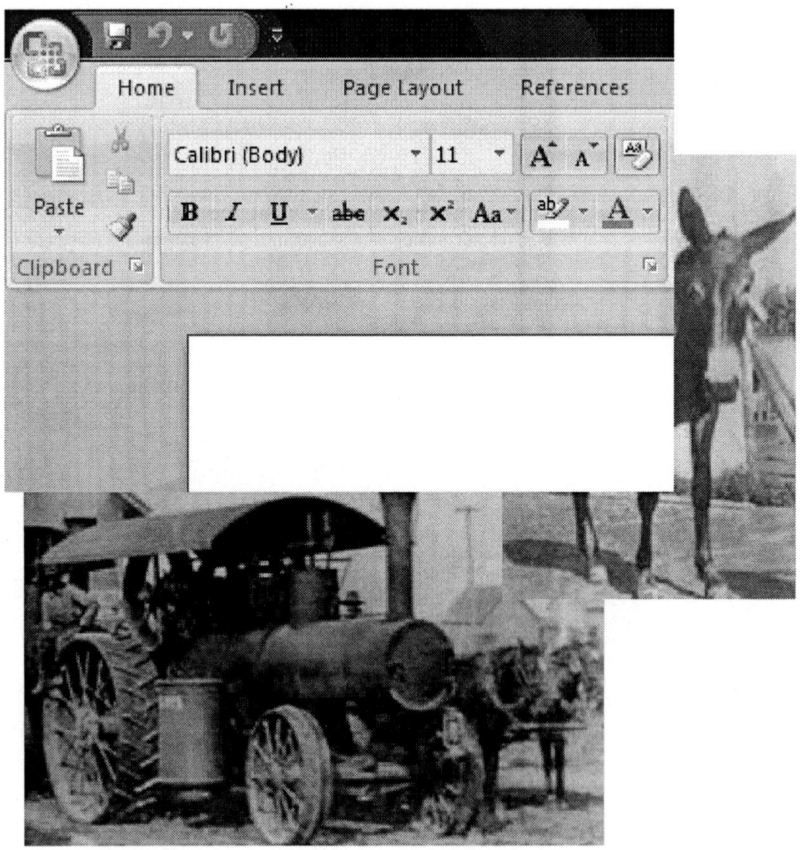

Take One

Home ->Font ->Color

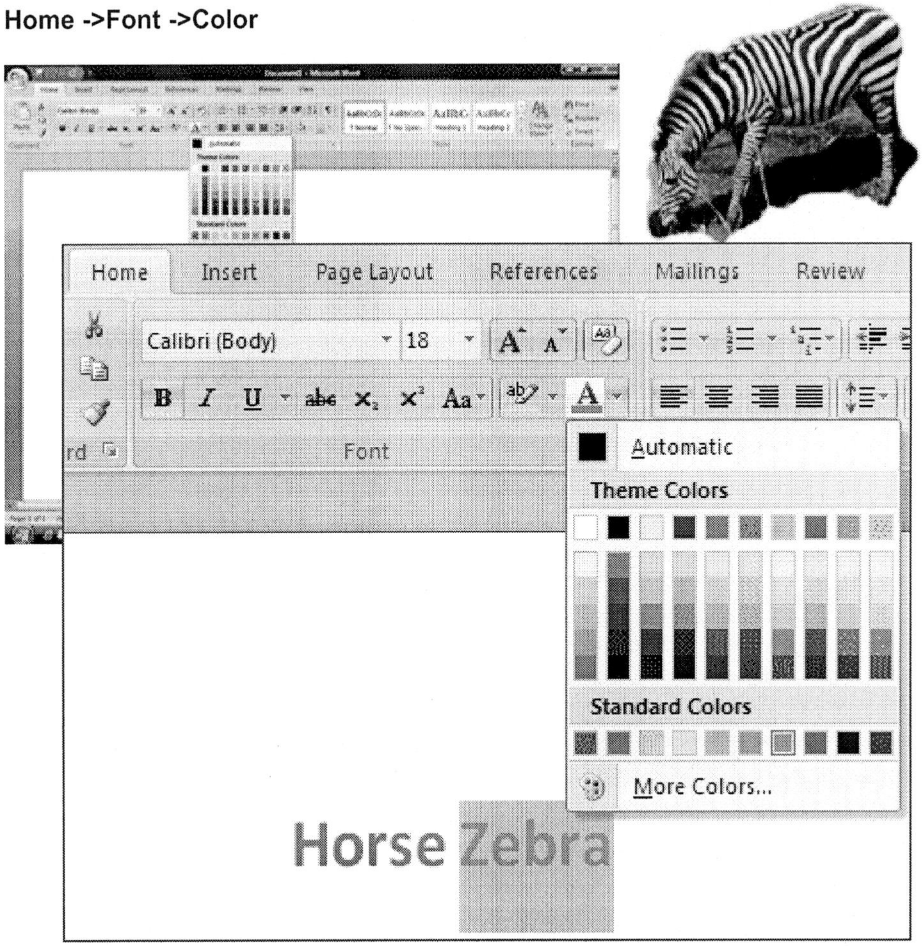

Format Text

Moving text and graphics from one page to another, or from one document to another, is a fundamental computer task. This lesson begins with text.

Try It: Format the Font
Type: Horse
Type: Zebra

Double click Horse to select it
Go to **Home ->Font**
Select 18 pt, Bold, Red

Double click Zebra to select it
Go to **Home ->Font**
Select 18 pt, Bold, Blue

Memo to self:
Nothing happens in a computer until you select it first.

Microsoft Word 2007 Exam 77-601 Topic: 2. Formatting Content
2.1. Format text and paragraphs
2.1.3. Format characters: Change font colors

Home -> Copy

Copy and Paste

Question: If you format the text big, bold and colorful, will Microsoft Word retain that formatting when you copy and paste the text?

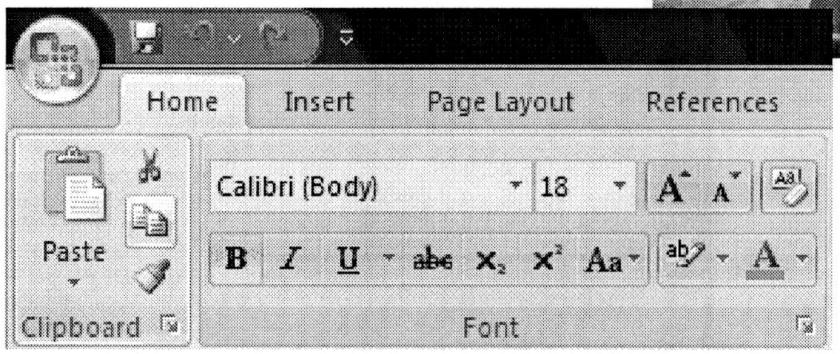

1. Try it: Copy the Text

Double click Horse to select it

Go to **Home ->Copy**

The Copy command looks like two sheets of paper. Look for it in the Clipboard group.

Paste the Text

Click your cursor after the word Zebra

Go to **Home ->Paste**

Paste the word Horse five more times.

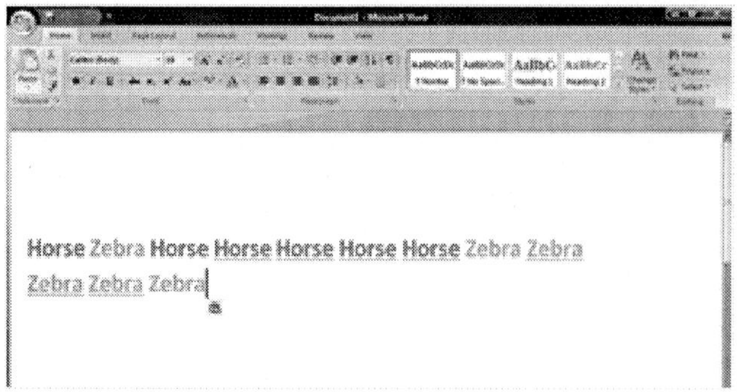

2. Repeat the practice

Double click Zebra to select it

Go to **Home ->Copy**

Place your cursor after the last word

Go to **Home ->Paste**

Paste the word Zebra five more times.

Microsoft Word 2007 -> Home

Drag and Drop Text

Our goal is to edit the words so that they go, “Horse Zebra Horse Zebra.” The action is called Drag and Drop.

Try it: Drag and Drop the Text

Double click Horse to select it
Click and HOLD your left mouse button The cursor changes from an I-beam to a white headed arrow with a box under it.

Drag and Drop the word horse between two zebras. You can see a little vertical line that follows your mouse as you move it. That’s the insertion point. No surprises. When you release your mouse that’s where the word you’re holding will be dropped.

Play with it a little to get used to drag and drop editing. It works. It’s fast. And when you get the hang of it, it’s fun.

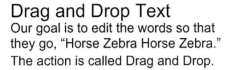

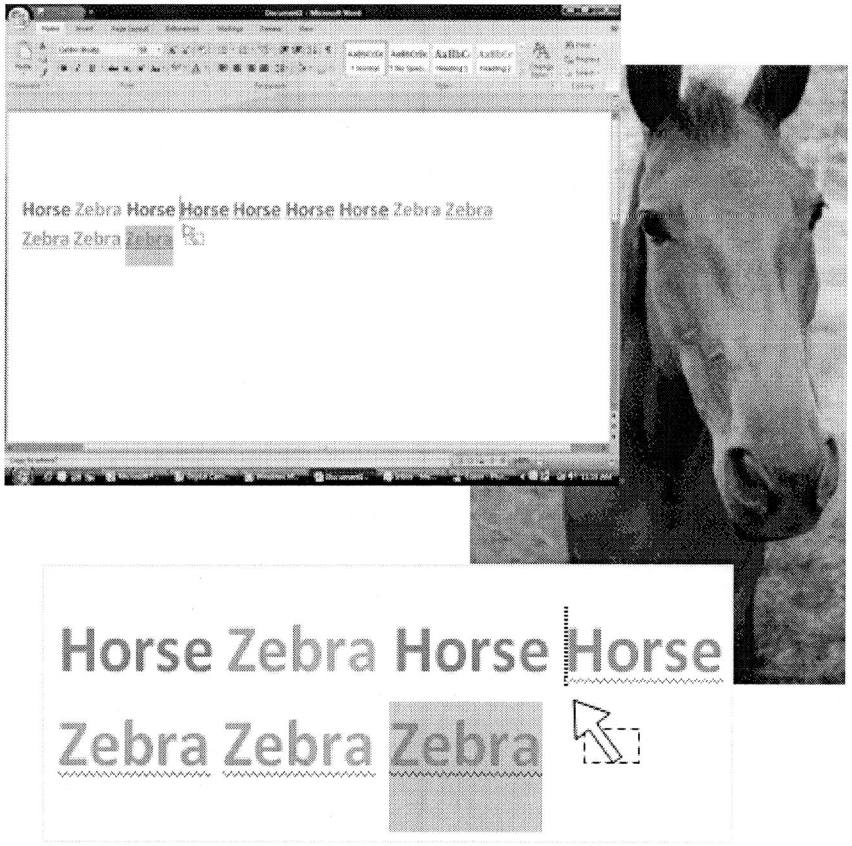

Insert -> ClipArt

Insert ClipArt

This exercise is called Horses and Zebras. The objective is to use the basic commands with pictures as well as text. So, you need horses and zebras.

Try it: Insert ClipArt

Go to the **Insert** tab

Find the Illustration group

Click on **ClipArt**

You will see a new **ClipArt Task bar** on the right side of your screen. What options are available?

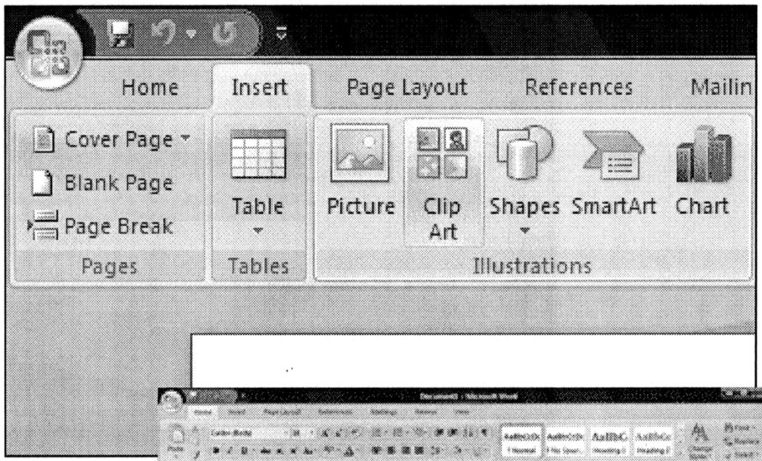

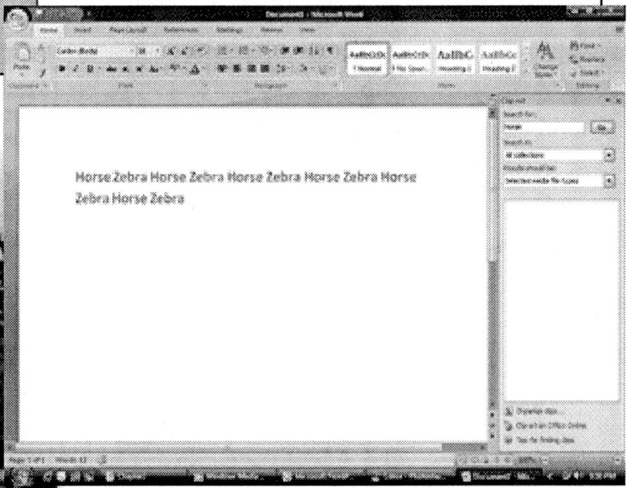

Hellllo ClipArt!

There are three criteria you can manage when you search for ClipArt: topic, location and media type.

When you look for photos, pictures, or video clips, you are not limited to the media available on your own drive. **Clip Art** can also search online for graphics, sounds and movies at the Microsoft Design Gallery and other web collections.

You can choose whether you want to look online from the options found under Search In.

You can also filter your search by checking whether you want some or all of the media types.

Insert -> ClipArt

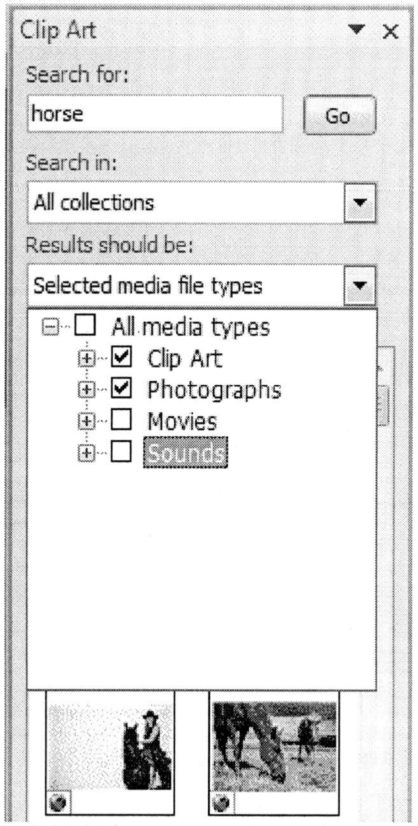

Insert -> ClipArt

Pick a Picture

Try it: Insert ClipArt

Type Horse for the Search text.
Click on **Go** to make it look for anything that matches your criteria.

The **ClipArt Gallery** will display a collection of horses.

Click on a picture that you want. It will be inserted into your Word document.

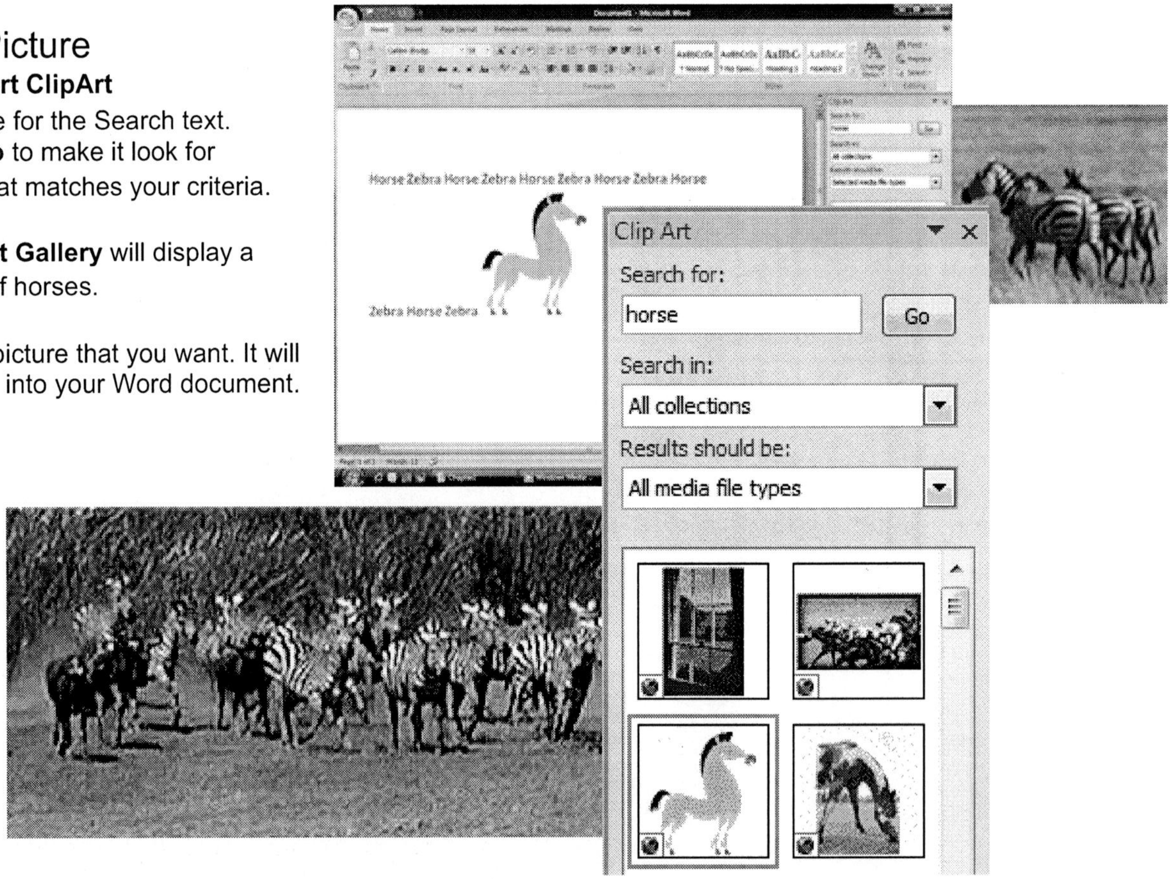

Picture Tools -> Format -> Position

Picture Tools

How does this picture interact with the text? Is it on the words, covering them up? Under the words? What is the **position**?

Try it: Edit the Picture Position.
Click once on the picture to **select** it.
Notice the new **Picture Tools.**
Select the **Format** tab.
Look for the **Arrange** group.
Try different **Positions** by sliding the mouse around.

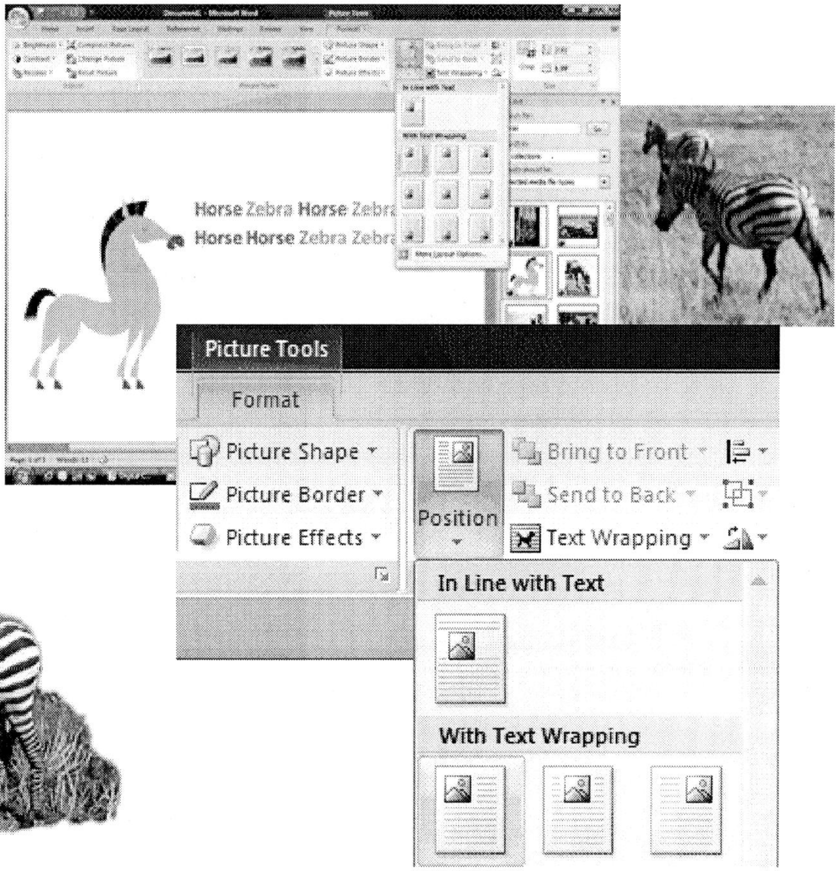

Resize the Picture

Resize the Picture

Click once on the picture to select it. See the little circles around the horse picture? Those are the **handles**. If you move your cursor to one the handles, the cursor will turn into a double-headed black arrow.

Watch your mouse:

A two-headed arrow means **Resize**.

A four-headed arrow means **Move**.

Try it: Resize the Picture

Select the picture

Run your cursor over any handle

Watch for a two-headed arrow

Click and Hold your left mouse button to **Resize** the picture.

Memo To Self: Pictures, photographs and even graphs are frame-based. The little handles that surround an object when you select it are the edges of the frame. Frames "float," they can be placed anywhere on the document.

Copy and Paste the Picture

Sometimes when you copy and paste a picture, the new images is placed EXACTLY on top of the original image with computer precision. It will be very difficult to see that you do, indeed, have one picture on another.

Watch your mouse:

A two- headed arrow means **Resize**.

A four-headed arrow means **Move**.

Try it: Copy and Paste the Picture

Click once to **select** the picture

Go to **Home ->Copy**

Go to **Home ->Paste**

Try it: Move the Picture

Select the new picture you just pasted

Watch for a four-headed arrow

Click and Hold your left mouse button to **Move** the picture to a new position.

Repeat the work with a Zebra picture.

Home -> Copy

Home -> Paste

Move the Picture

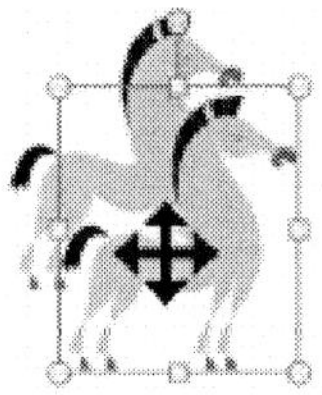

Test Yourself

Take One

Text and Pictures

This lesson focused on the different options available with the mouse. The exercise began with text and the work was repeated with graphics. Please close this file.You do NOT have to save this practice document when you are prompted.

You done good.
You get the cookie.

1. Which group includes the commands Copy and Paste?
a. Home
b. Clipboard
c. Copy and Paste
Tip: Beginning Word, page 44

2. What media types are available in Microsoft's Clip Art Gallery?
(Select all that apply)
a. Clip Art
b. Photographs
c. Sounds
d. Movies
Tip: Beginning Word, page 47

Downloads

Level 2 Horse and Zebra 2007
Level 2 Horse and Zebra 97-2003
Graphic Files
Horse, Horse2, Zebra, Zebra 2

Assessment

There is no **Skill Test** for this lesson. You can continue to the next lesson if you wish.

Mice and Men

Mice and Men

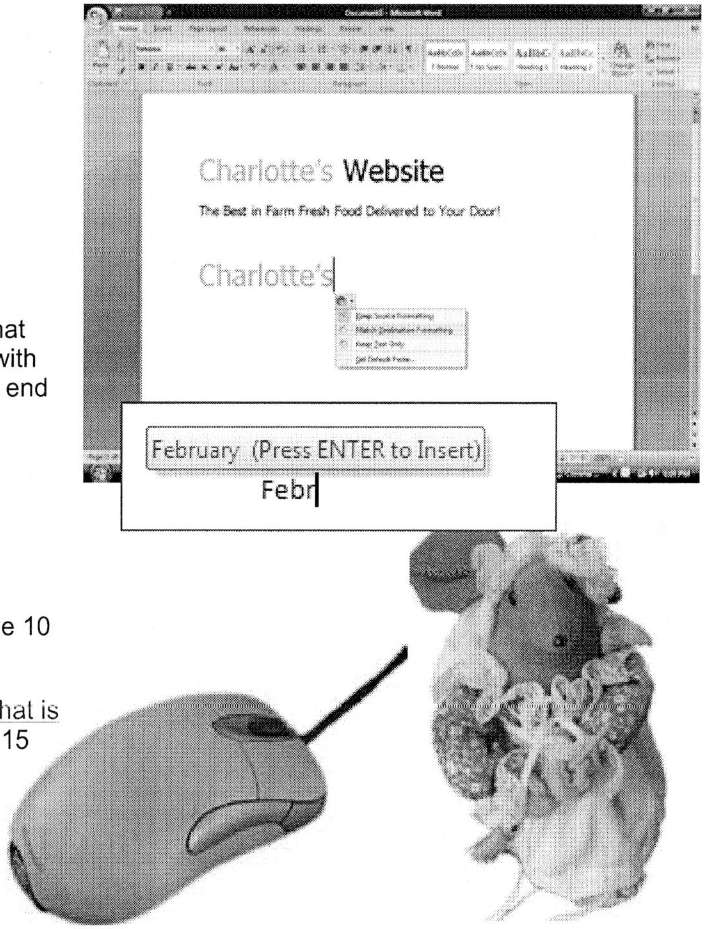

Click Here to Get Started
Sample Files

Beginning Word

Lesson Objectives: This lesson focuses on the options that are available with the contextual menus that are available with the Left and Right mouse buttons in Microsoft Word. At the end of this lesson you will be able to:

Identify the Mini Toolbar page 4

Demonstrate the right-click Paste options page 5

Show how to correct Spelling and Grammar errors page 7

Identify and use AutoText and AutoCorrect page 9

Locate additional Word Options under the Office menu page 10

Demonstrate how to Save your document page 13

Use Save As Office 97-2003 to make a copy of your work that is compatible with previous versions of Microsoft Office page 15

Mice and Men

You have to talk with your computer. In one way or form, you need to use an interface to change your words into actions in the digital world. Microsoft Office can be used for dictation, but the majority of our computer interaction is with a **mouse**. In this lesson, the focus will be on what you can think and do with the mouse. **Start** the **Program** Microsoft **Word**.

What do you see from the top of the screen? Is there a Title Bar that says Microsoft Word? Yes.

Is there a **Home** Ribbon with the **Clipboard, Font and Paragraph** Groups? Yes.

If your screen looks similar to the example on this page, then you are ready to get started.

Home -> Font

Begin with the text

Enter the following sample text.

1. Enter the name
Type: Charlotte's Website
Select the text
Go **Home** and select Tahoma, 36 pt

2. Enter the marketing text
Type: The Best in Farm Fresh Food Delivered to Your Door!
Select the text
Go **Home** and select Tahoma, 16 pt

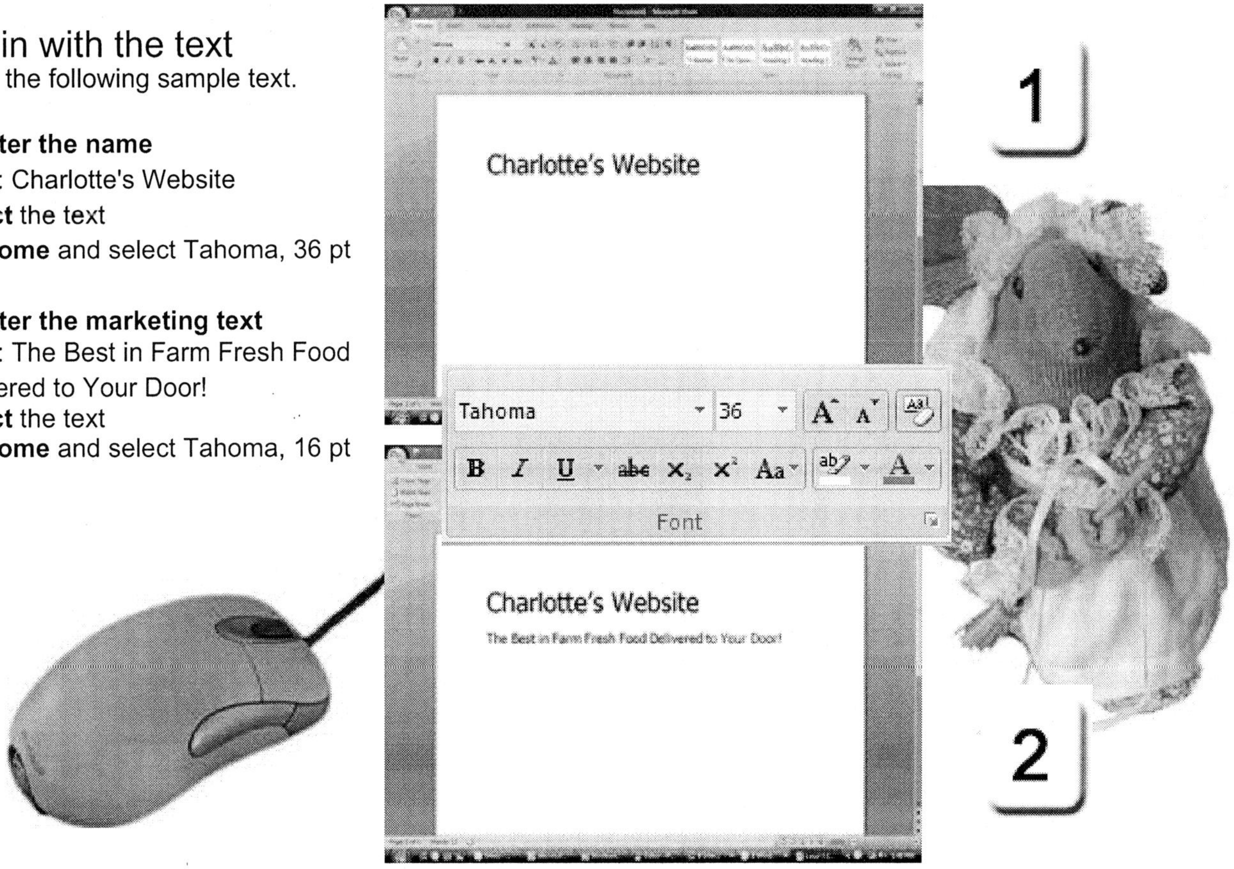

The Mini Toolbar

When you select text, you may see a **Mini Toolbar.** It looks transparent until you rub your mouse over any of the formatting commands.

You can use this toolbar to edit the Font, text size, alignment, color, indents and even bullets.

Try it: Format the text
Select the text: Charlotte's
Look for the **Mini Toolbar**
Change the size and color

Home -> Font

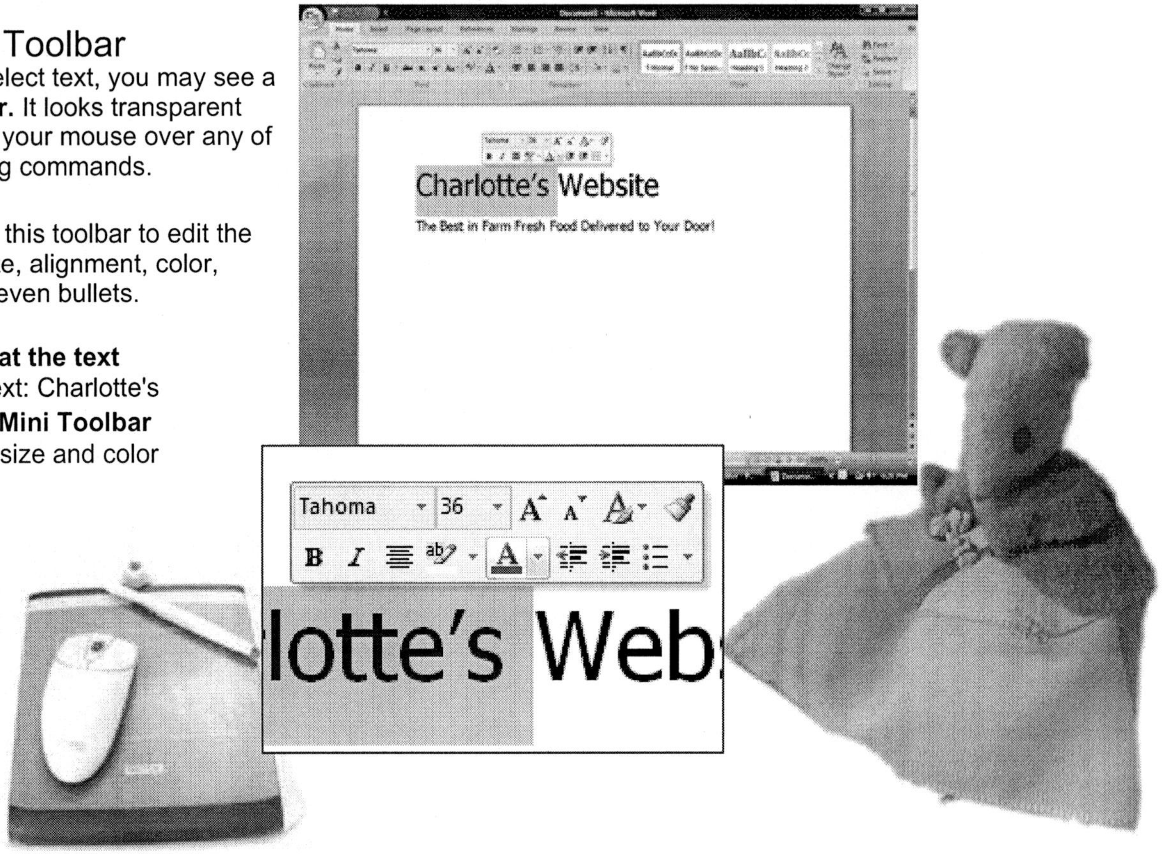

Home ->Paste

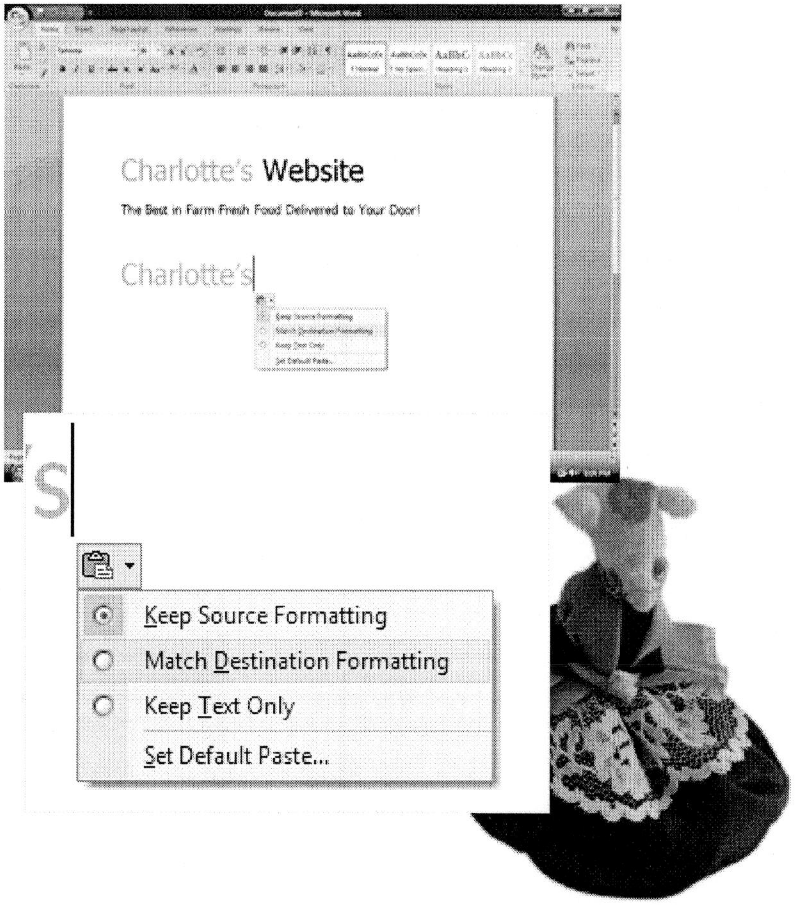

Paste Options

When you copy and paste text, you may see a small **Clipboard** pop up. There are three formatting choices in these **Paste Options**.

Try it: Paste Options
Select the text: Charlotte's
Copy and Paste the text.
Look for the Clipboard pop up.

Keep Source Formatting:
In the example on this page, the word "Charlotte" would paste with all of the formatting that you copied: big, bold and colorful.

Match Destination Formatting:
The word "Charlotte" would paste with the default font, size and color for Microsoft Word: Calibri, 11 pt, black.

Keep Text Only:
This option strips away any formatting you may have copied. This is a good way to simplify the text you copy from the Internet.

Paste One, Paste All

The **Clipboard** is one of the oldest helper applications in the computer. You can use the Clipboard to copy and paste a paragraph or a picture from one program to another.

Try it: Use the Clipboard
Double click the word: Website.
Go to **Home -> Clipboard**. Click on the small option arrow to the right of the label for the Clipboard.

What Do You See? A new pane will open on the left side of Microsoft Word. Each item that you copied will be listed here. Click on any item to paste it. You can also **Paste All**, if you wish.

Home ->Clipboard

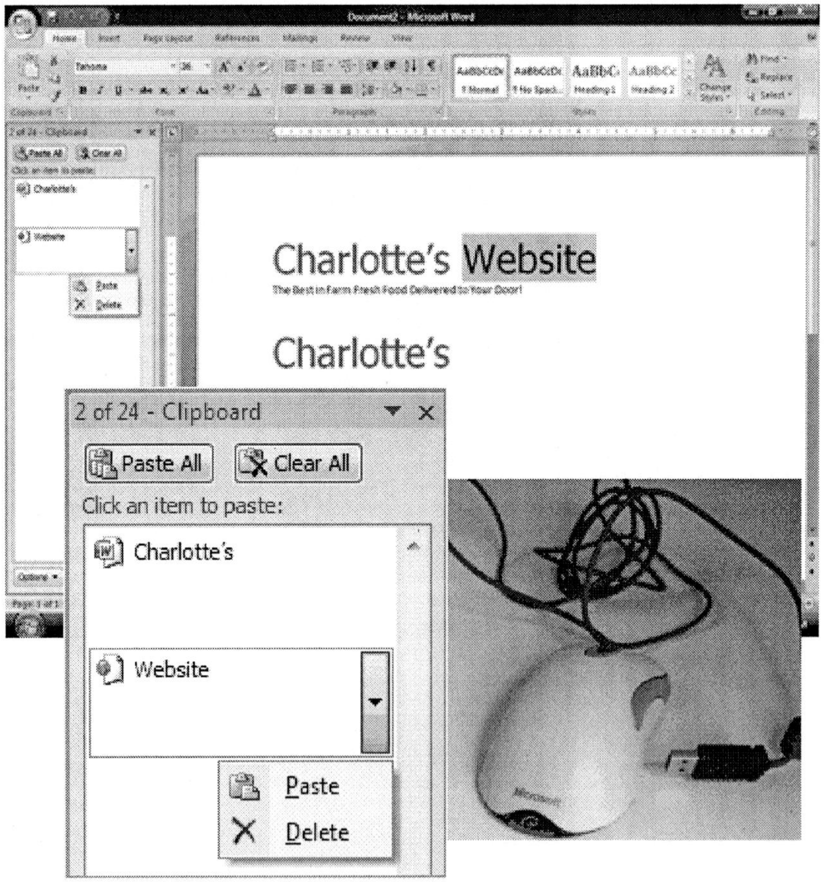

Review ->Spelling and Grammar

Spelling and Grammar

Microsoft Office includes a dictionary. It compares your spelling and grammar as you are typing. If your words do not match ones in the dictionary, you will see a red wavy line. That's the **Spell Checker**.

If your sentence has a structural problem with the grammar, say the subject and the verb do not match, then you will see a green wavy line. That's the **Grammar Checker.**

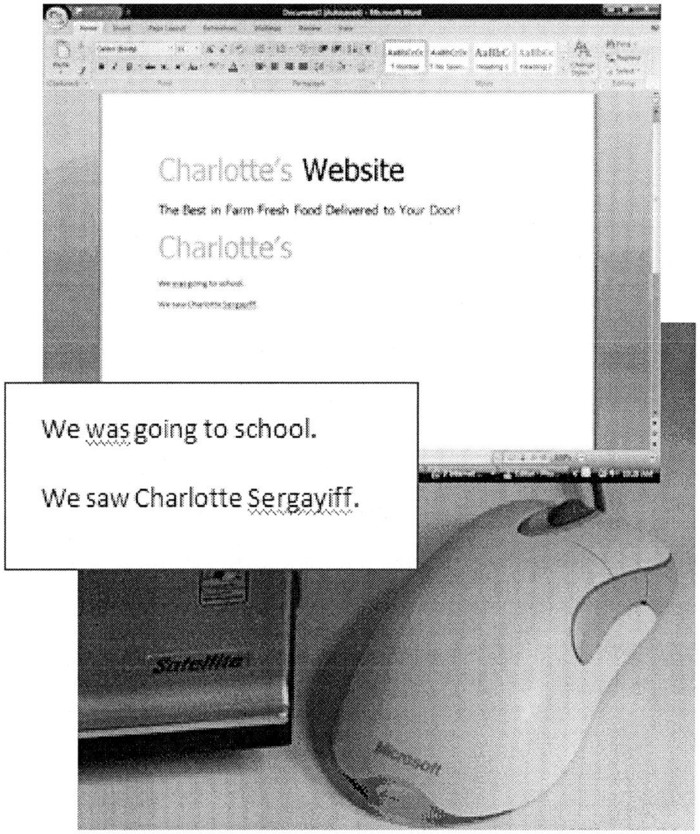

Try it: Spell and Grammar Check

Type the following two sentences:
We was going to school.
We saw Charlotte Sergayiff.

Please include all of the punctuation!

Click **Enter** on your keyboard. What do you see?

Spelling and Grammar

Everything in a computer has a left and a right mouse action. What does the left mouse button do?

What mouse button to you use to **click** on Start?
Say left.
What mouse button do you use to **double click**?
Say left.
What mouse button do you use to **drag and drop**?
Say left.

Left gives you action: Select, open, and move.
Right gives you options: A short list of choices.

Try it: Right click the Grammar
Select **were**, the correct grammar from the list of options, by clicking on it with your Left mouse.

Try it: Right click the Spell Check
Sometimes, you will see a few suggestions from the dictionary. If the name is spelled correctly but does not appear on the list of options, click on **Add to dictionary** with your Left mouse.

Review ->Spelling and Grammar

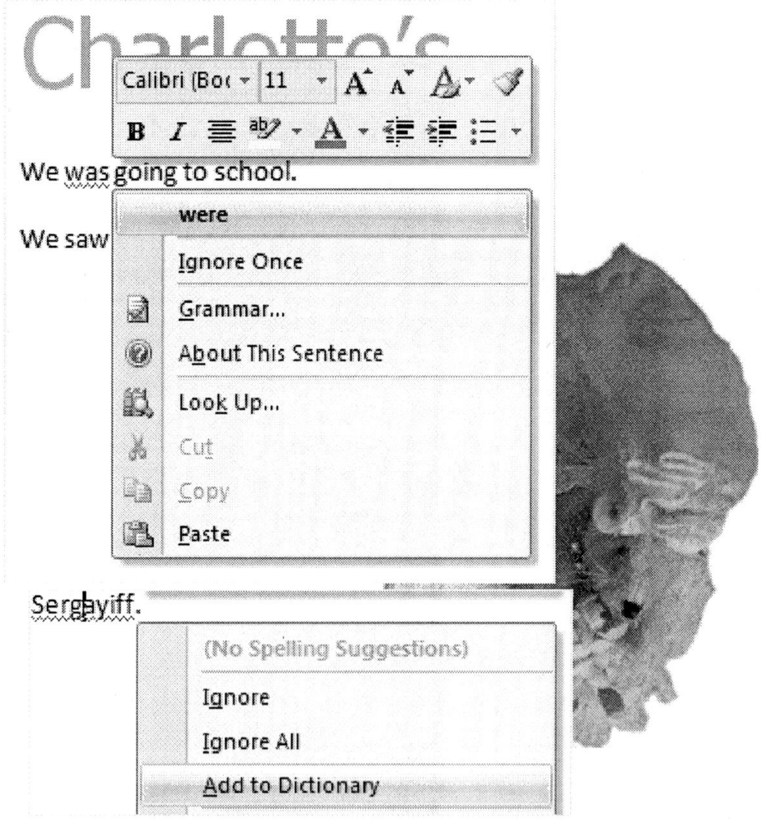

AutoText, AutoCorrect and Other Fancy Mouse Options

AutoText: Microsoft Office also notices common words and offers to fill them in for you. The most obvious example is the names of the months.

Try This: Play With the AutoText

What do you see as you type the word February?

AutoText also inserts formatted symbols. If you type (tm), Microsoft Office will format the text as a superscript. If you type :) Microsoft Office will substitute a smiley face. It is the Computer Mama's understanding that the symbol next to the smiley face means SmartAss, which would be accurate for her.

AutoCorrect: Microsoft Office automatically corrects a long, long list of the most common spelling mistakes. Prove it to yourself. Type: hte and watch the letters when you press the space bar.

Office ->Word Options

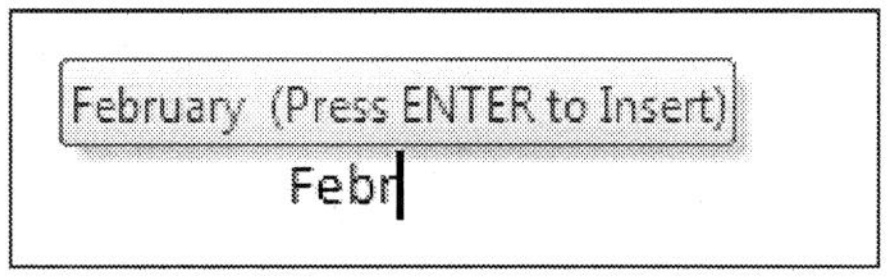

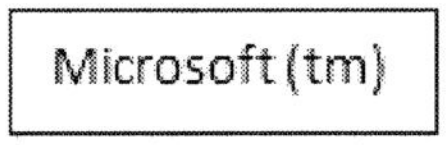

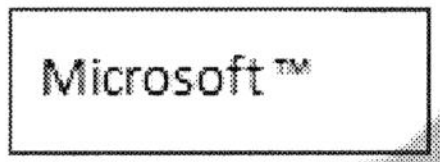

Word Options

This is an effective piece of software. But where are all of these dictionary words and AutoText entries?

Try it: Review the Word Options

Go to **Office** in the upper left corner

Click on **Word Options**

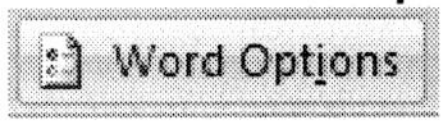

You will be taken to a collection of **Word Options**:

Popular

Display

Proofing

Save

Advanced

Select **Proofing** and review the defaults.

Office ->Word Options

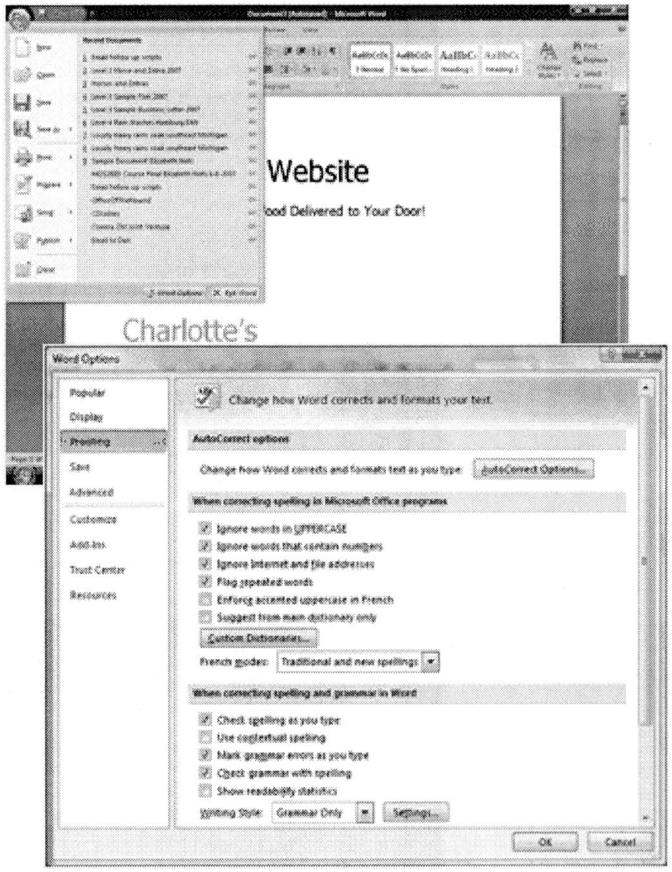

Proofing Options

The **Proofing** options are found by going to Office and selecting Word Options from the bottom of the menu. Go a little further and review the following.

AutoCorrect ignores words with:
UPPERCASE, numbers, Internet addresses.

AutoCorrect underlines common mistakes:
Repeated words, capitalization, grammar.

AutoText can also be a productivity tool.
Say you worked for the Livingston Regional MTEC. This name is a mouthful, especially if you are required to type it many times in each customer support letter.

You can add a phrase to AutoText and have Microsoft Office spell it out for you automatically.

Office ->Word Options -> Proofing ->AutoCorrect

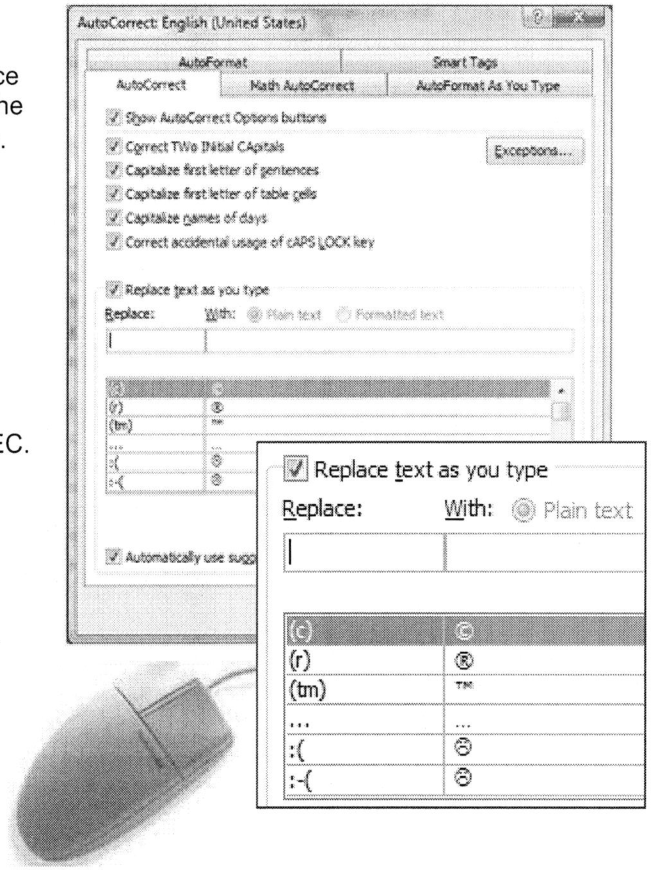

Office ->Word Options -> Proofing ->AutoCorrect

Math AutoCorrect Options

Many professions use mathematical symbols in reports and documentation. Until Office 2007, you had to **Insert** a **Symbol** and select from a set of custom Fonts named Dingbats or Symbols.

Math AutoCorrect has an extensive list of replacements for math, chemistry and engineering. This list is adaptable, too. You can add or edit your own entries, same as with AutoCorrect.

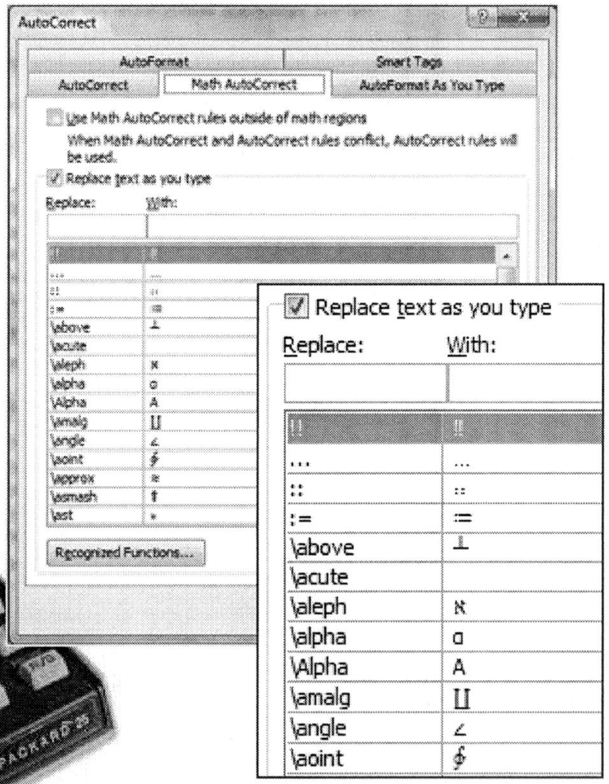

Save Your Document

There are three parts to saving a file:

1. Where are you saving it?
2. What are you naming it?
3. What are you doing? SAVE!

These are the steps to save your work and find a folder to keep it in.

Try This: Save a document
Start Microsoft Word
Type your name
Go to **Office ->Save.**

Keep going...

Office ->Save

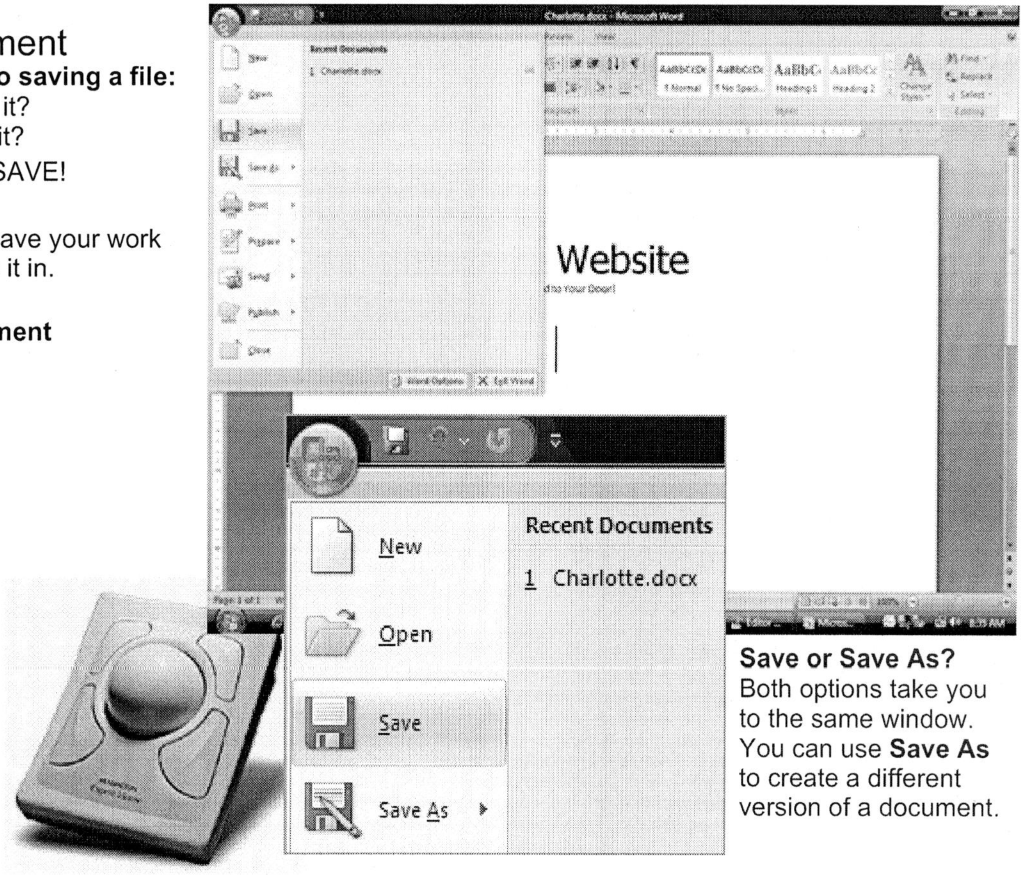

Save or Save As?
Both options take you to the same window. You can use **Save As** to create a different version of a document.

Microsoft Word 2007 Exam 77-601 Topic: 6. Sharing and Securing Content
6.1. Prepare documents for sharing
6.1.1. Save to appropriate formats

Office ->Save

Save Options

1. Where Are You Saving It?
By default, Microsoft Office offers your **Document** folder on your computer's hard drive as a place to save it. You can use the Documents folder if you wish.

2. What Are You Naming It?
Type the File Name: Charlotte

What Do You See?
The new **file type** in Microsoft Word 2007 is the **docx** format. The docx format looks like a Word document, but it is the new international Open Office standard.

3. What Are You Doing?
Click on **Save.**

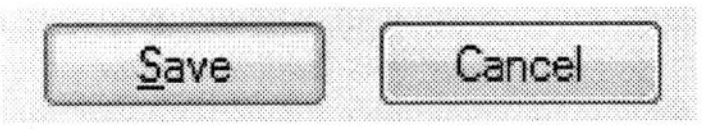

When you click on **Save**, your document will be named, date stamped, and stored in the Documents folder.

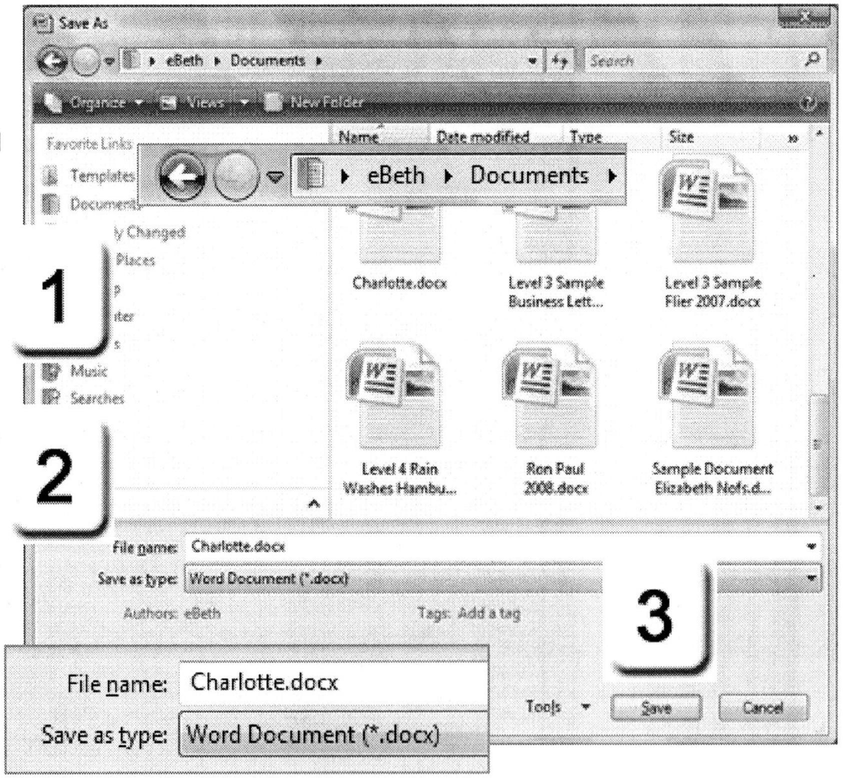

Save As Office 97-2003

When you save a new document in Microsoft Word 2007, you should be aware that this is a **new file format** and that companies with a previous version of Microsoft Word might not be able to open or edit your work.

Here are the steps you can take to create a **copy** in Word 97-2003.

Try it: Save As Previous Version

Go to **Office ->New**.

Create a blank document.

Type: Dr. Green

Go to **Office ->Save As**.

Select: **Word 97-2003 Document**.

Keep going...

Office ->Save As -> Word 97-2003

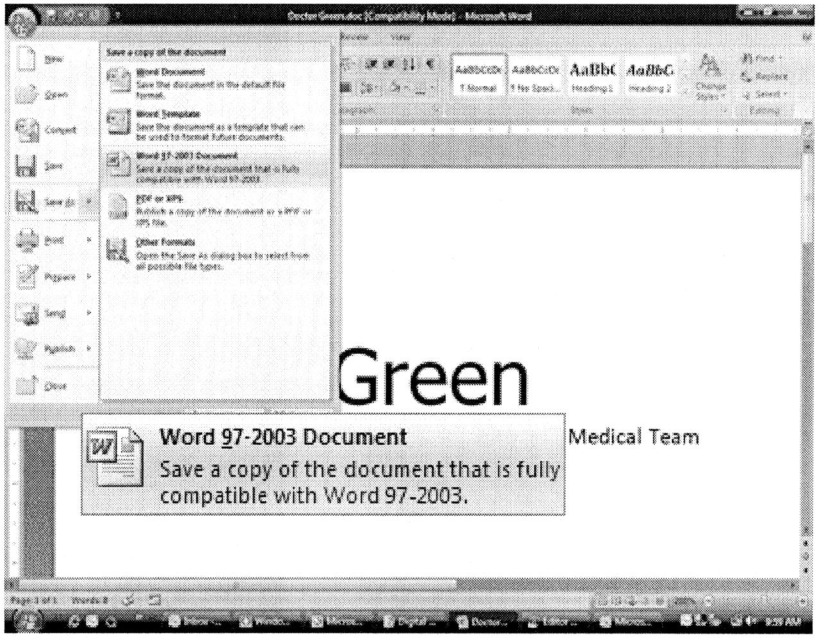

Memo to Self: If you can't remember when you had to consider which version of Microsoft Office you are using, then you are not as old as the Computer Mama. The last time anyone had to ask, "Which Word?" was in 1997.

Microsoft Word 2007 Exam 77-601 Topic: 6. Sharing and Securing Content

6.1. Prepare documents for sharing

6.1.1. Save to appropriate formats: Save as a .doc, .docx, .xps, .docm, .or dotx, file

Take Two

Save As Office 97-2003

1. Where Are You Saving It?
You can use the Documents folder if you wish.

2. What Are You Naming It?
Type the File Name: Doctor Green

What Do You See?
The file type is Microsoft Word 97-2003. It is the **doc** format. This is the file type that is compatible with most businesses, schools and government departments.

3. What Are You Doing?
Click on **Save.**

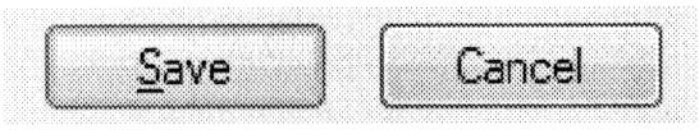

When you click on **Save**, your document will be named, date stamped, and stored in the Documents folder.

Office ->Save As -> Word 97-2003

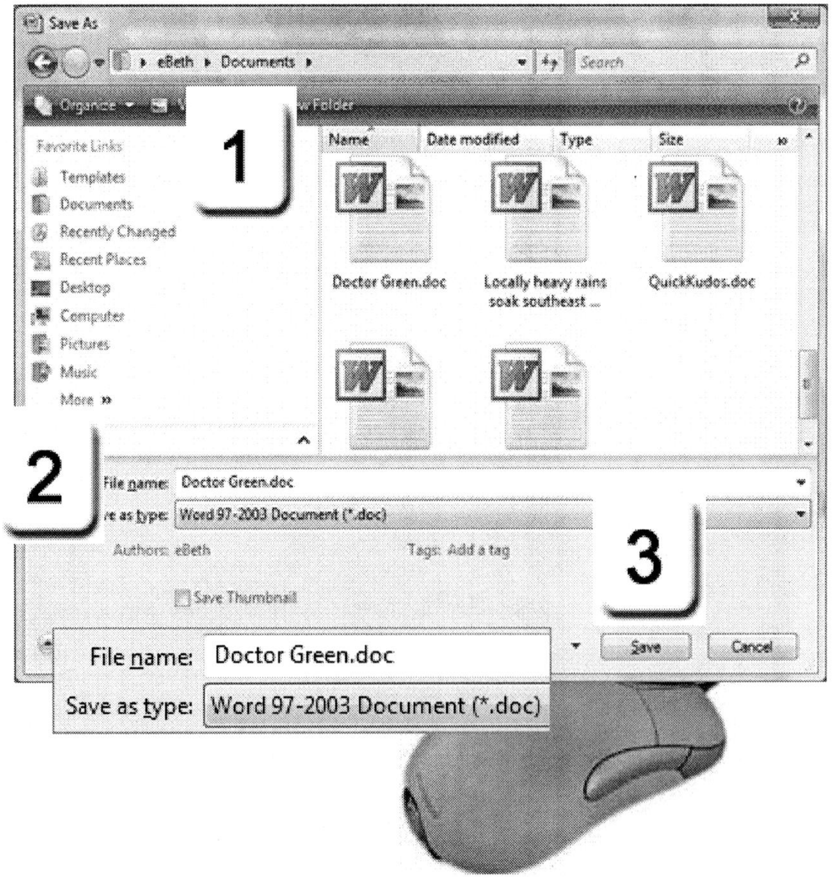

Microsoft Word 2007 Exam 77-601 Topic: 6. Sharing and Securing Content
6.1. Prepare documents for sharing
6.1.1. Save to appropriate formats: Save as a .doc, .docx, .xps, .docm, .or dotx, file

Office ->Save As

Save As a Different Copy

Say you needed to consult with two doctors: Dr. Green and Dr. Cooper. Say the letter you typed to Dr. Green will be the same as letter to Dr. Cooper except you need to edit the name and address.

You can use **Save As** to create **another copy, or version** of your letter.

Try it: Use Save As to Make A Copy
Open the file saved as: Doctor Green. Edit the text: Delete the name Green and type Cooper.

Go to **Office -> Save As**.
Where are you saving it: Documents.
What is the File Name: Doctor Cooper.
What are you going to do? Click Save.

Memo to Self: When you use Save As to create another copy of a Word 97-2003 document, Word uses the same file type as the original.

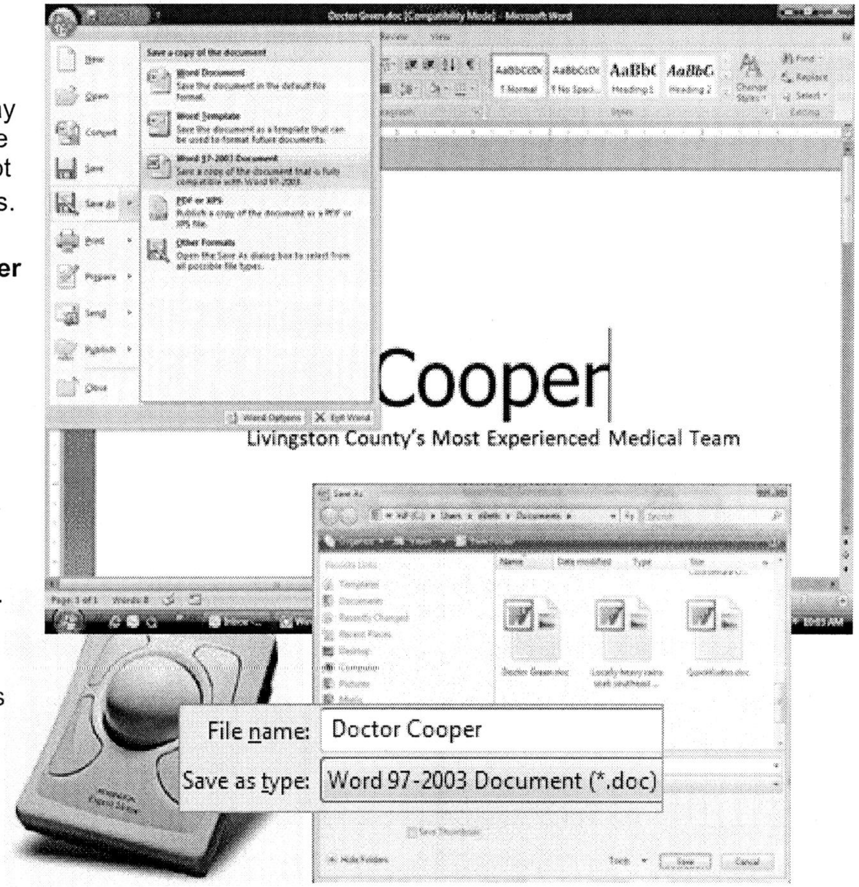

Microsoft Word 2007 Exam 77-601 Topic: 6. Sharing and Securing Content
6.1. Prepare documents for sharing
6.1.1. Save to appropriate formats: Save as a .doc, .docx, .xps, .docm, .or dotx, file

Office ->Word Options -> Save

Where Did You Save It?

All of the **Microsoft Office** programs save files in the Documents folder. You can change the default location if you wish. This setting is under the **Office** button, with the AutoText options and other tidbits.

Review the Default Location for Saving Your Documents
Go to **Office ->Word Options**.
Select the category: Save.

What Do You See? You can use the **Browse** button to find and select a different folder to be the default location when you Save a file.

Please **Cancel** out of these Word Options without changing the default file location.

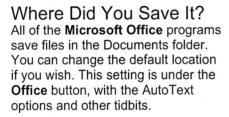

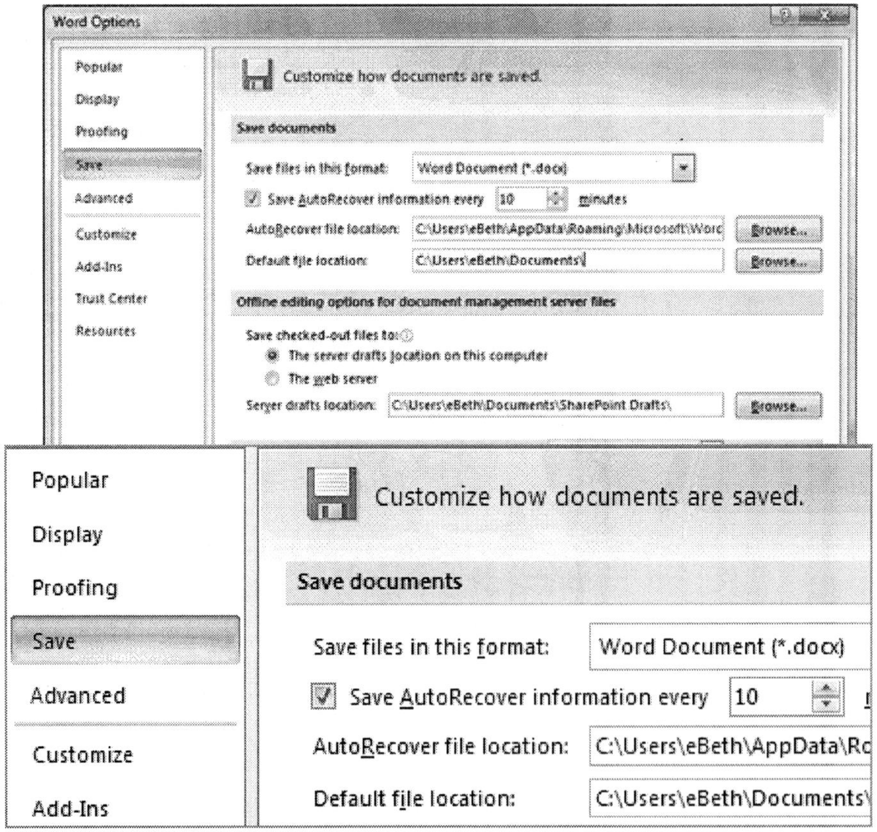

Take Two

Start ->Documents

Beginning Word

These pages demonstrated many of the options you might see as you learn Microsoft Word 2007. It is important to consider who has upgraded to the new Word 2007 format, and how you can use Save As to create a version that is compatible with Word 97-2003.

Finally, this lesson introduced the concept of different versions. This page shows two copies of the same letter, one for each consultant in the Word 97-2003 document (doc) format..

Well, you done good.
You can have two cookies.

Test Yourself

1. When does the Mini Toolbar appear?
a. When text is selected
b. When you double click on the ribbon
c. When you start typing
Tip: Beginning Word, page 56

2. The clipboard holds items copied:
a. Just from the current program you're using
b. Only from Microsoft programs
c. From any program on your computer that allows copying
Tip: Beginning Word, page 58

3. What does Paste All do?
a. Pastes all items from the clipboard into the current document
b. Pastes all items copied that day into the current document
c. Pastes only the last copied item into document
Tip: Beginning Word, page 58

4. Which is correct about using the Grammar Checker?
(Select all that apply)
a. Grammar errors are underlined with a green wavy line
b. Select and use right click on a grammar error to open the grammar checking menu
c. Grammar checking works similarly to Spell Check
Tip: Beginning Word, page 59, 60

5. Where is the Options button ?
a. On the Home Ribbon
b. In the menu under the Office Symbol
c. Under the Tools Menu
Tip: Beginning Word, page 62

6. Word 2007 includes Math Autocorrect, which replaces certain command with symbols.
a. TRUE
b. FALSE
Tip: Beginning Word, page 64

7. It is possible to create personalize AutoCorrect commands.
a. TRUE
b. FALSE
Tip: Beginning Word, page 63

8. Microsoft Word 2007 uses a new file format that is different from previous versions of Microsoft Word.
a. TRUE
b. FALSE
Tip: Beginning Word, page 66

Downloads

Biz1, Biz2, Biz3, Flag1, Flag2, Flag3,

Assessment

There is no **Skill Test** for this lesson. You can continue to the next lesson if you wish.

Page 1 2 3 4 5 6 7 8 9 10 11 12 13 14 15 16 17 18 19 20 21 22 23 24 25 26 27 28

WYSIWYG

First Impressions

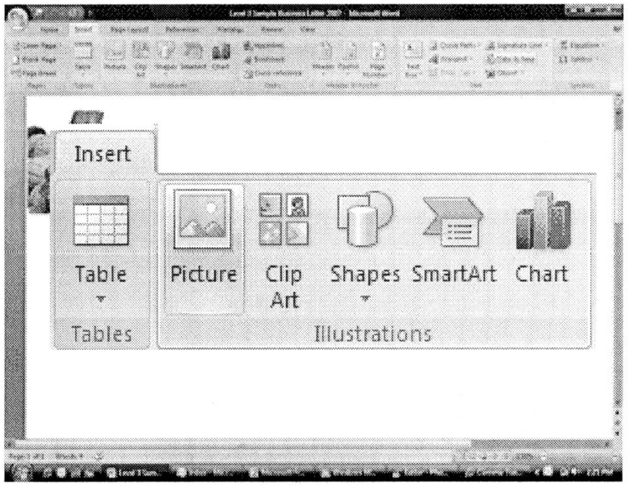

Click Here to Get Started
Sample Files

Beginning Word

Lesson Objectives: Learn how to create your own stationery and save it as a Template. This demonstration shows options for working with Visual Content and inserting illustrations, Organizing Content with Quick Parts, and saving your work in another format. In this lesson you will:

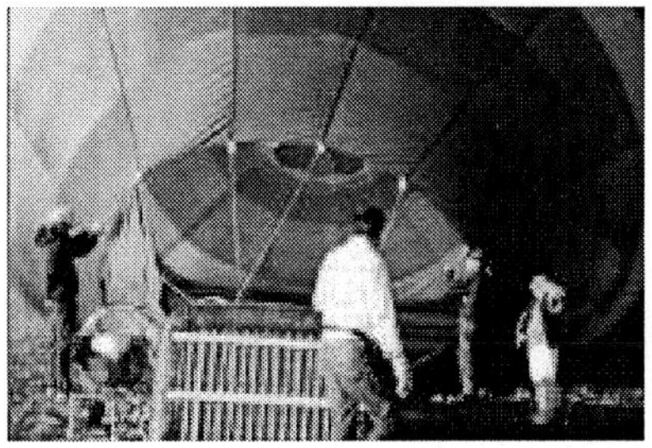

Create Business Stationery

The objective of this example is to use the formatting options in Microsoft Word to create business stationery. The logo and type setting on business stationery is sometimes called the Corporate Stripe. Formatting helps your company documentation look consistent. It also identifies a corporate brand. Ok, Go **Start** the **Program Microsoft Word**.

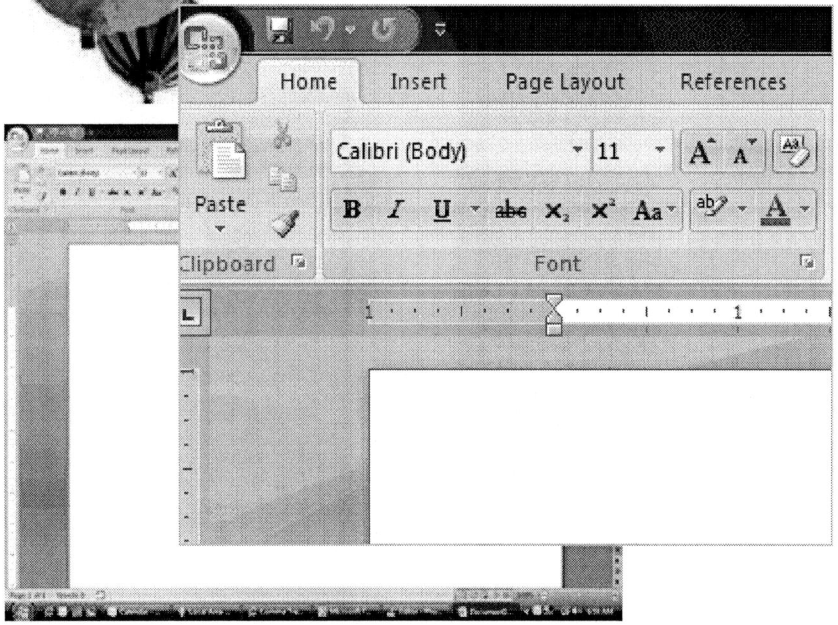

What do you see, from the top of the screen? Is there a **Title Bar** that says Microsoft Word? Yes.

Is there a **Home** Ribbon with the **Clipboard, Font and Paragraph** Groups? Yes.

If your screen looks similar to the example on this page, then you are ready to get started.

Begin the Document

The purpose of this exercise is to use the tools and options to create a professional business letter. Through out this work, you can practice with the sample company, or develop your own. We will also insert a picture for our company logo.

1. Type the name and address

Please type:

Charlotte's Web Site

123 Main Street
Brighton MI 48116.

2. Select the text

Nothing happens in a computer until you select it, first. Try selecting the type backwards. For some reason, it has always been easier in Windows to highlight backwards.

Here are the steps: Click your cursor by the 6 in the Zip code and hold your mouse down as you drag backwards to the C in Charlotte's.

Home -> Font

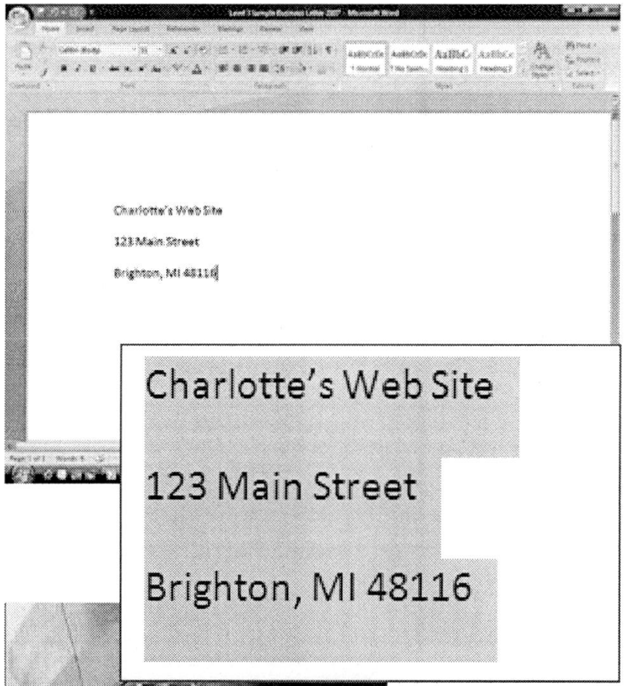

1

2

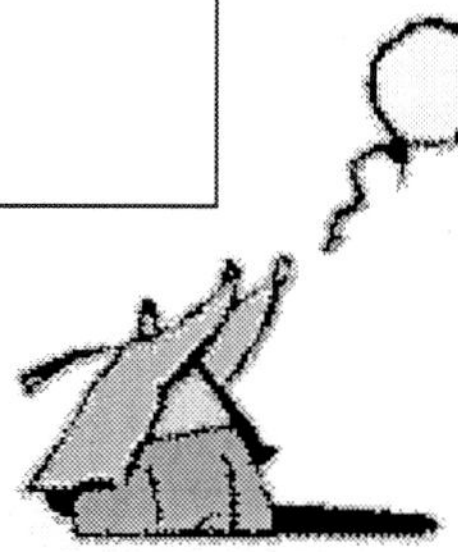

Take One

Format the Type

Calibri is the default type face—or font—for Microsoft Word 2007. The company name and address should be special. It is supposed to call attention to itself. The name can be differentiated with big, bold type.

3. Select a Different Font
Select the name and address.
Go to **Home->Font**
Click on the little down arrow by the words Calibri and select Tahoma from the list. It is a distinctive, professional font.

4. Center the Paragraph
Select the name and address
Go to **Home ->Paragraph**
Click on the **Center** button.

Not sure which button is which? If you hold the mouse over the button, a label will pop up and tell you what it is. These labels are called ToolTips.

Home -> Paragraph

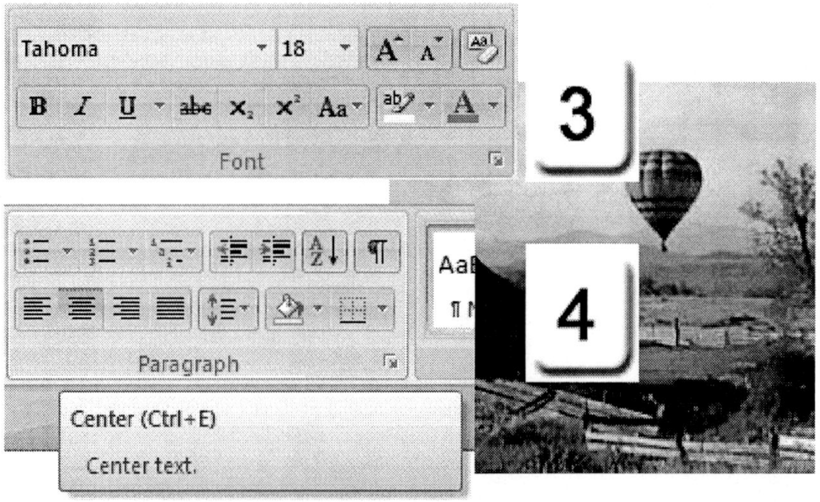

Take One

5. Emphasize the Name
Select Charlotte's Website
Change the size of the business name by clicking on the down arrow to the right of the "12."
Choose "18" from the list.

Now, the company name is bigger and bolder than the address.

Home -> Font

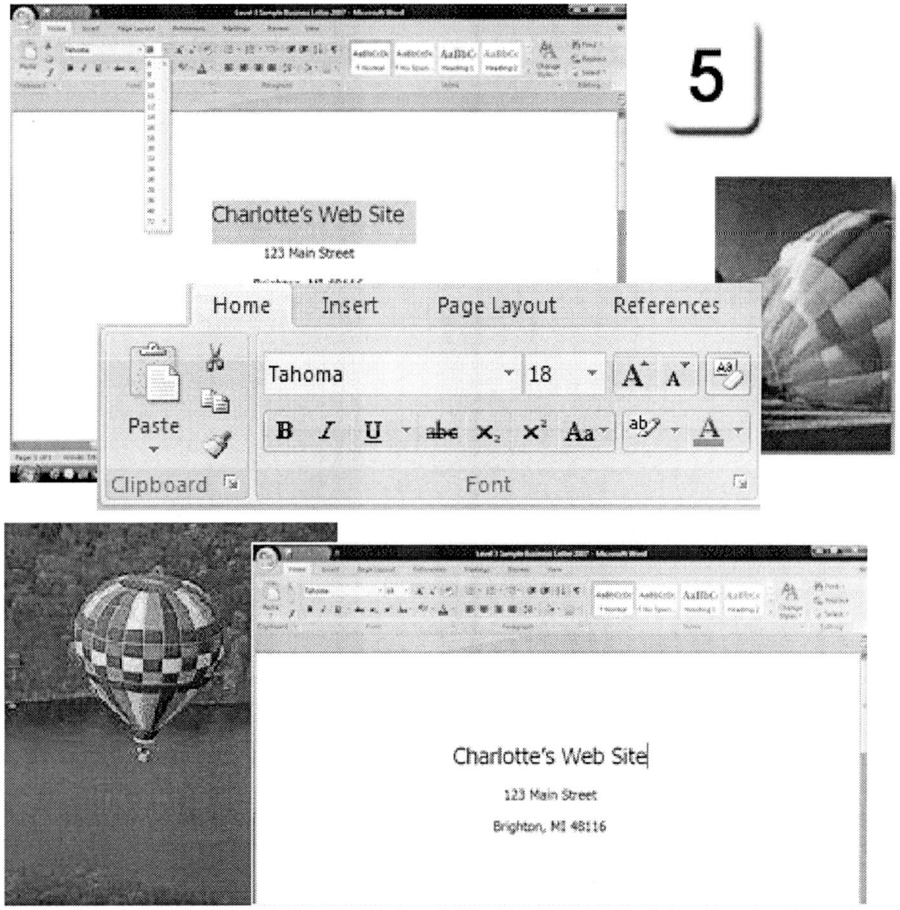

Microsoft Word 2007 Exam 77-601 Topic: 2. Formatting Content
2.1. Format text and paragraphs
2.1.3. Format characters: Change font size

Take One

Add a Company Logo

A company or department's logo is a brand that identifies your team work. It is important not only for sales, but for your own team experience: We belong.

There are some **sample files** that you can copy to your **Documents** folder, or you can use your own graphics.

6. Insert a Picture

Go to the **Insert** Ribbon.

Click on **Picture**.

You will be asked to locate your pictures. The next steps show you how to find the files in the **Documents** folder.

Insert -> Picture -> From File

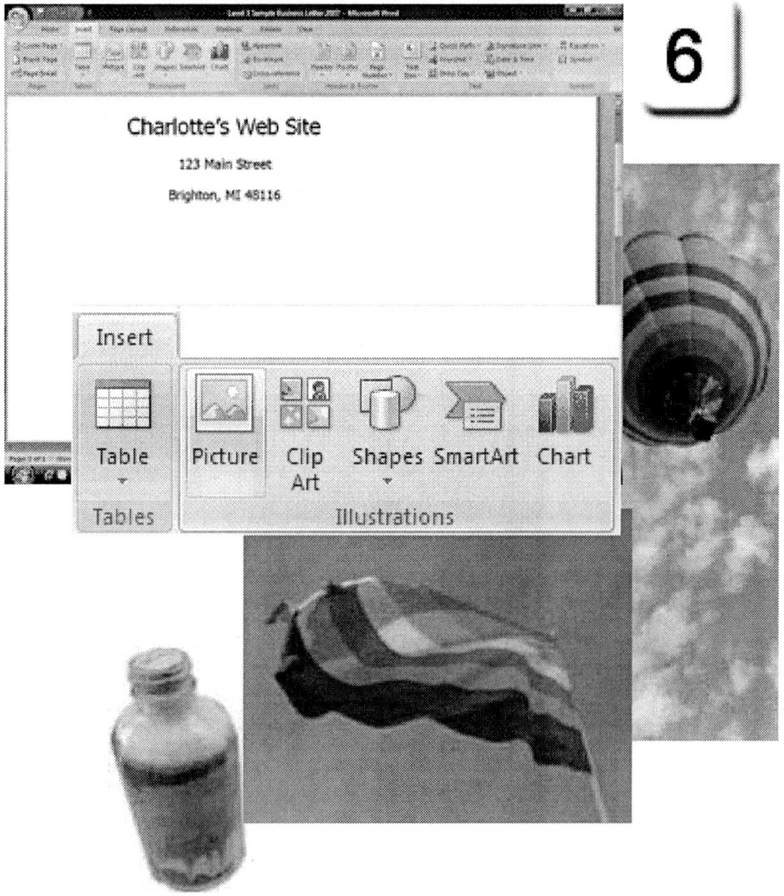

Find a Picture

The **Pictures** and the **Documents** folder can be found in your **User Name** folder. In this example, the User name is eBeth. Some camera programs download your pictures to their own folders.

7. Locate and Select a Picture
Look in the **Documents** folder

Double click one of the sample pictures to insert into your document.

You can also select the picture and then click on the **Insert** button in the lower right hand corner.

Insert -> Picture -> From File

7

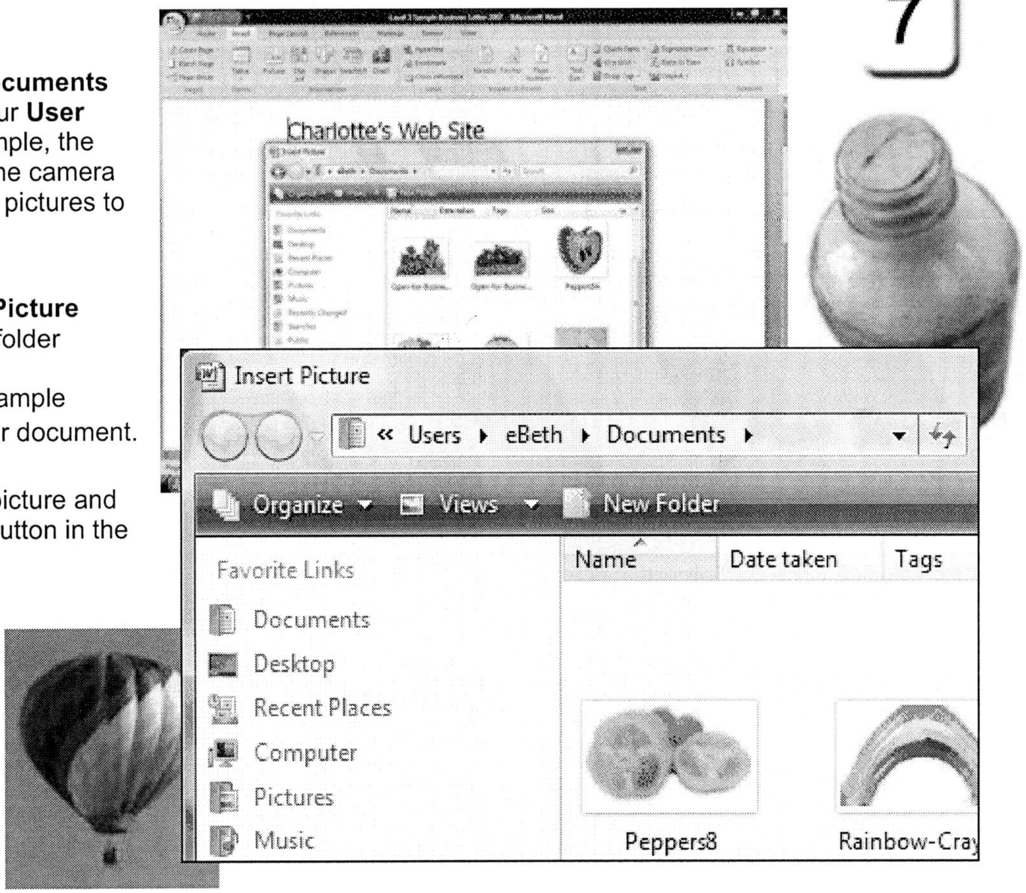

Take One

Format the Picture

The picture was inserted into the document as **In Line With Text**. This is the default setting for how pictures interact with text.

In effect, the picture is anchored to the line, same as any word. The picture will be easier to work with if you change the **Text Wrapping.**

Tight means that the text will flow, or wrap, around the picture. This makes it easy to move the picture like a sticky-note.

8. Change the Text Wrap to Tight

Click once on the picture to select it

Go to the **Format** Ribbon

Find **Arrange**

Select **Text Wrapping ->Tight**

Picture Tools -> Format ->Arrange ->Text Wrapping

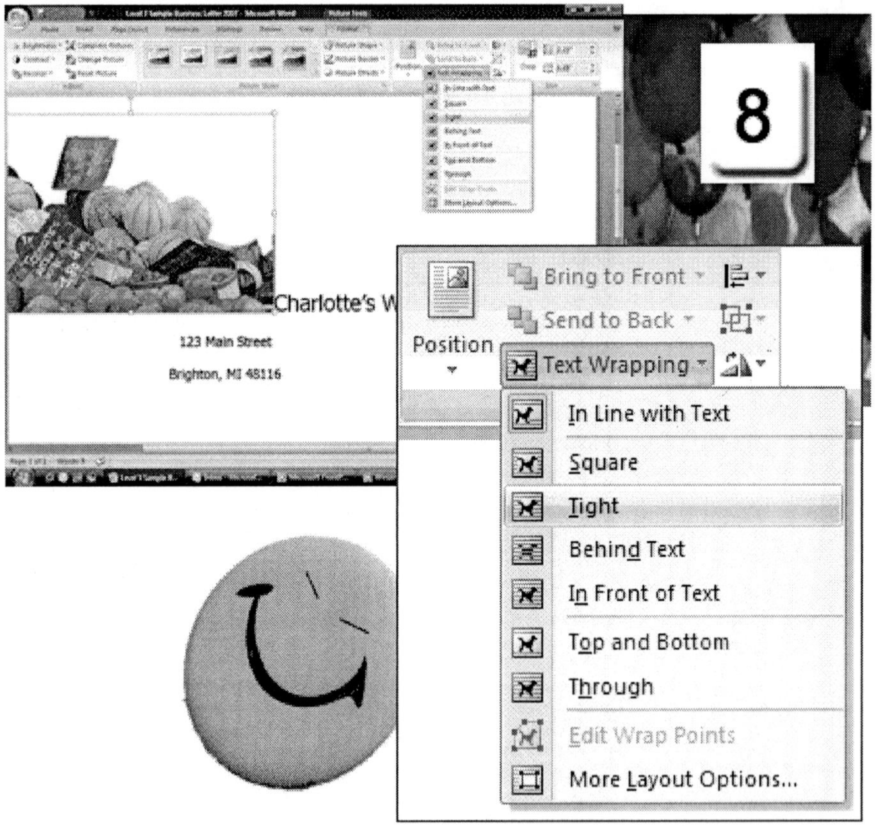

Insert ->Date & Time

Text Blocks

Microsoft Word 2007 has **Building Blocks** that help you compose a professional letter. These steps show how to **Insert** the **Date and Time.**

Before You Begin
Place your cursor after the zip code and press **Enter** a couple of times to create a few blank lines. You may need to go to the **Home** Ribbon and format the text to be 11 pt Tahoma, aligned left.

9. Insert the Date and Time
Go to **Insert ->Date & Time.**
Select a **Medium** date format**.**

A **Medium** date format spells out the month, day and year. **(MM/DD/YYYY)**.
Some countries write the date as day, month, year. (DD/M/YY). Using the Medium format could minimize the confusion with the Short date format that uses only numbers.

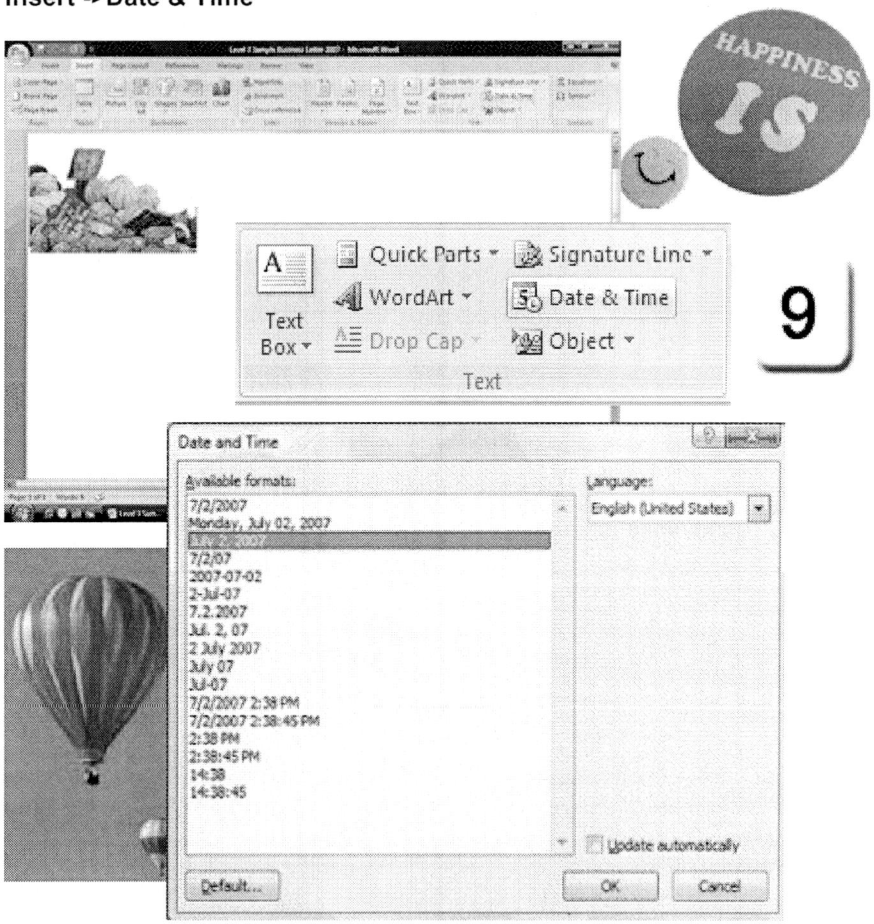

Microsoft Word 2007 Exam 77-601 Topic: 4. Organizing Content
4.1. Structure content by using Quick Parts
4.1.1. Insert building blocks in documents

Create a Sample Letter

Finish the exercise by typing a sample business letter.

Suggested text:
Dear Sir,

Thank you for your recent order from Charlotte's Website.

Sincerely...

Do This: Save your Letter
Go to **Office -> Save**.
File Name: Charlotte's Letterhead.

Office -> Save

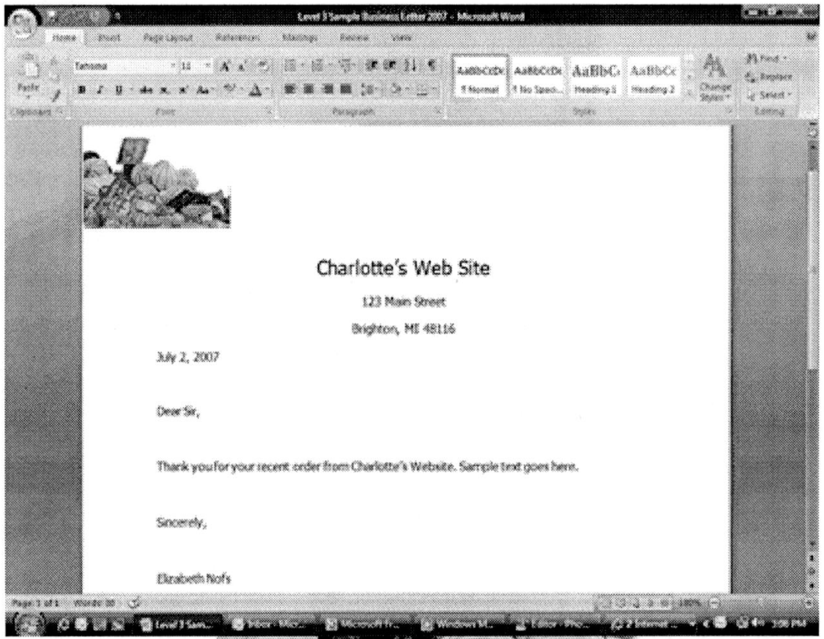

Quick Parts

Each company or department builds a collection of documents that identifies who they are when they communicate with their clients. These documents can include stationery, envelopes, fax cover pages and newsletter.

Many of these documents use the same elements that we designed into the letterhead: the company name, address, and logo.

Try This: Insert a Quick Part

Open the sample letterhead your created for Charlotte's Web Site.

Select the text: Charlotte's Web Site.
Go to **Insert ->Quick Parts**.
Choose **Document Property ->Company**.
Keep going...

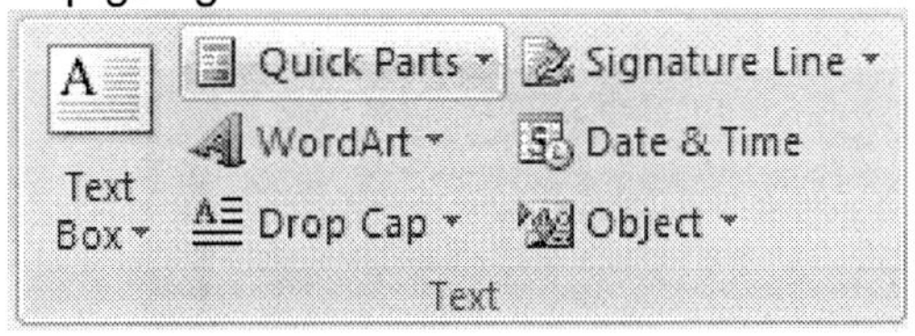

Insert ->Quick Parts ->Document Property

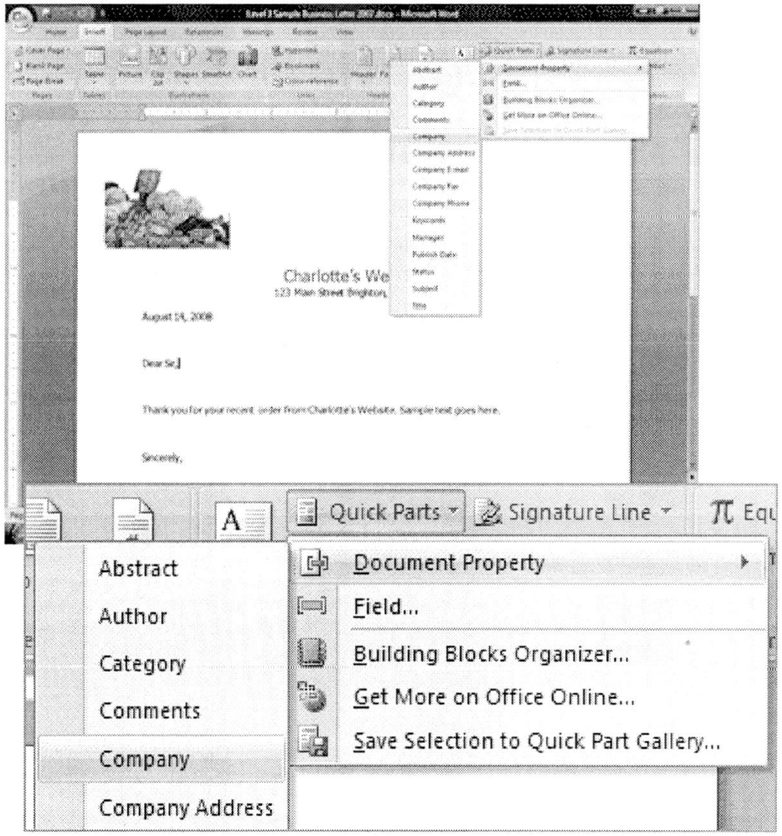

Take One

Insert ->Quick Parts ->Document Property ->Company

Edit the Building Block

What Do You See? When you create a new Quick Part, you should see a little **Building Block** with the label: Company.

Type the Company Name: Charlotte's Website.

Format the Text: Go to the Font tools on the Home menu and select:
Font: Tahoma.
Size: 18 Pt
Color: Green.

Good Practice: Use the **Document Property** to insert and edit the Company Address, E-mail, Fax, Phone number or other contact data.

Save your updates.

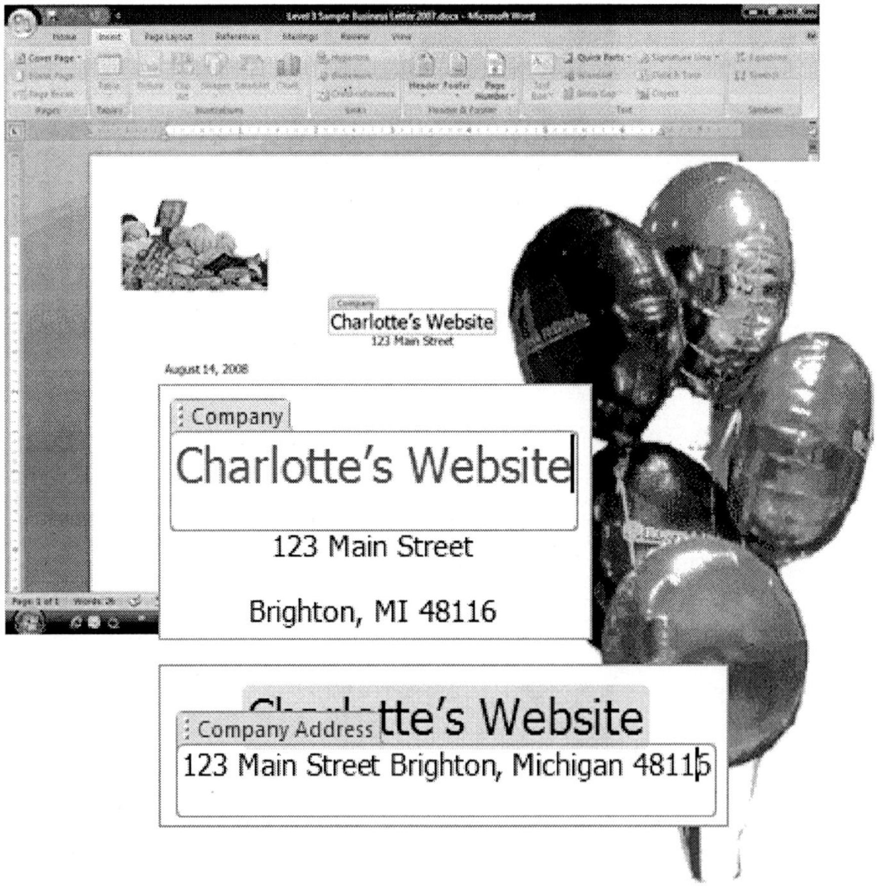

Save Selection

Say you already had a logo that you wanted to use again and again in other documents. You can save that logo to the **Quick Part Gallery.**

Try This: Save Selection
Click once on the logo to select it.
Go to **Insert ->Quick Parts**.
Choose: **Save Selection to Quick Part Gallery**

You will be prompted to add this picture as a **Building Block** in the Quick Part Gallery.

Keep going...the steps are on the next page. ;-)

Insert ->Quick Parts -> Save Selection to Quick Part Gallery

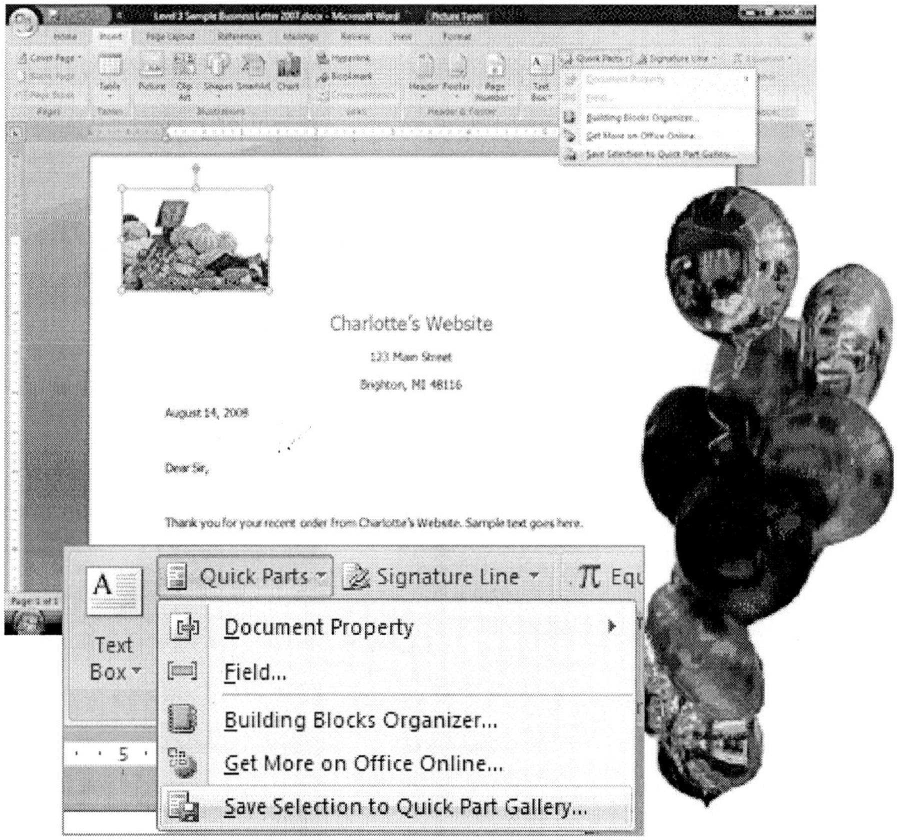

Microsoft Word 2007 Exam 77-601 Topic: 4. Organizing Content
4.1. Structure content by using Quick Parts
4.1.2. Save frequently used data as building blocks: Save company names or logos as building blocks

Take One

Make a Building Block

What Do You See? The Building Blocks are part of an extensive table, or database. There are six **properties** you can edit to identify and organize your Building Blocks. By default, a new Building Block will be added to the Quick Parts Gallery.

Try This: Edit the Building Block
Name: Charlotte's Logo.
Gallery: Quick Parts.
Category: General.

Accept the defaults for the following:
Save in: Building Blocks.dotx.
Options: Insert content only.

Memo to Self: Look in the **Gallery** and you will see a long list of document objects The Complete Computer Guides will show how to add and modify several Building Blocks in subsequent lessons.

Insert ->Quick Parts -> Save Selection to Quick Part Gallery

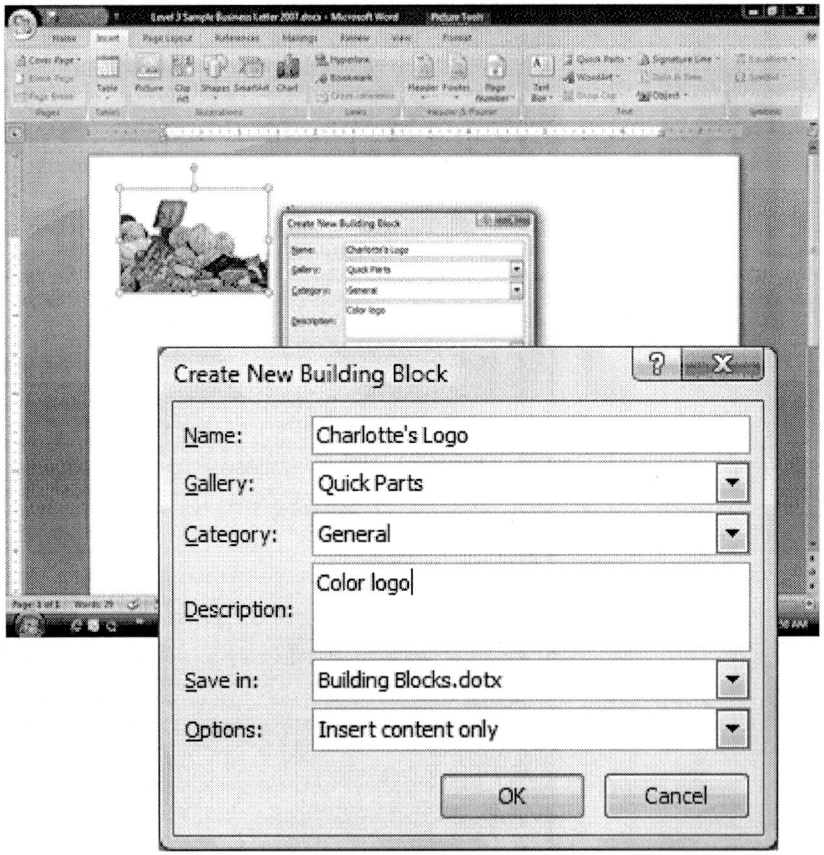

Microsoft Word 2007 Exam 77-601 Topic: 4. Organizing Content
4.1. Structure content by using Quick Parts
4.1.1. Insert building blocks in documents: Edit the properties of building block elements

Quick Parts Gallery

What Was The Result? When you go to **Insert ->Quick Parts**, you should see Charlotte's Logo in the Gallery. You can add your own Building Blocks and have them available whenever you need them.

Good Practice: Use the Gallery
Go to **Office ->New**.
Select a **Blank** document.
Go to **Insert ->Quick Parts**.
You should be able to add the new logo as well as the Company Name and Address.

Very good. If you made a practice document, you can can close it without saving. The examples on the next pages continues with the sample document for Charlotte's Web Site.

Insert ->Quick Parts

Microsoft Word 2007 Exam 77-601 Topic: 4. Organizing Content
4.1. Structure content by using Quick Parts
4.1.2. Save frequently used data as building blocks: Save company names or logos as building blocks

Take One

Insert ->Quick Parts -> Building Block Organizer

The Building Block Organizer

Microsoft Word 2007 gathers all of the Building Blocks into an Organizer. You can sort the Building Blocks by Name, Gallery, or Category.

Try It: Organize Building Blocks
Go to **Insert ->Quick Parts**.
Select **Building Block Organizer**.

What Do You See?
At the top of the list of Building Blocks are the column headers. Click on Name, and the list will be sorted alphabetically A-Z. Did you notice that many of the Names begin with the same template, such as Accent or Motion?

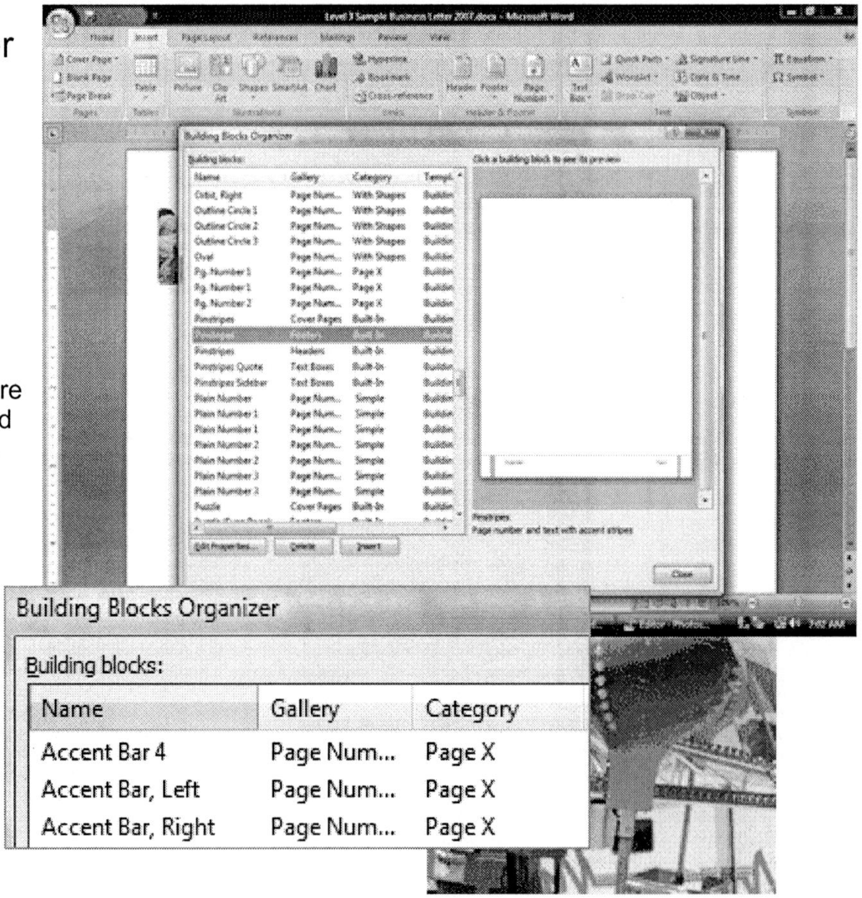

The Quick Parts Footer

Try This: Insert a Footer
Go to **Insert ->Quick Parts.**
Select **Building Block Organizer.**
Scroll down to Pinstripes and double click the **Footers**.

What Do You See? Look at the bottom of your document. Your cursor should be in a new footer at the bottom of your page.

Suggested Sample Type:
Charlotte's Web Site (810) 555-1212.

The footer is formatted with the font, size, color and alignment of the Pinstripe template. Can you select the type and use the tools on the **Home Ribbon** to edit the formatting?

Insert ->Quick Parts -> Building Block Organizer

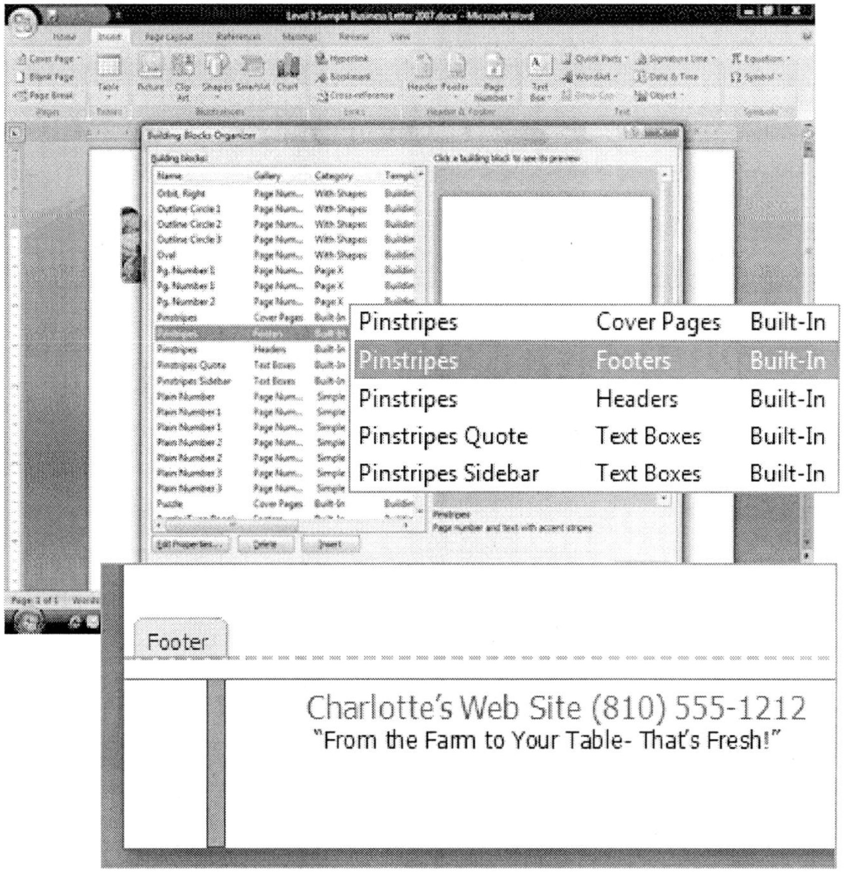

The Quick Parts Side Bar

A **Side Bar** is a column of additional information on the left or right of a document. One example of a side bar would be a formal letter that lists the executives and director's names.

Try This, Too: Add a Side Bar
Go to **Insert ->Quick Parts**.
Select **Building Block Organizer**.
Sort the Building Blocks by Category.
Find the Pinstripe Sidebar.
Click on **Insert**.

Add the following sample text:
Fresh Fruits
Apples
Avocados
Kiwis
Pears

Fresh Vegetables
Chives
Tomatoes

Save your changes.

Insert ->Quick Parts -> Building Block Organizer

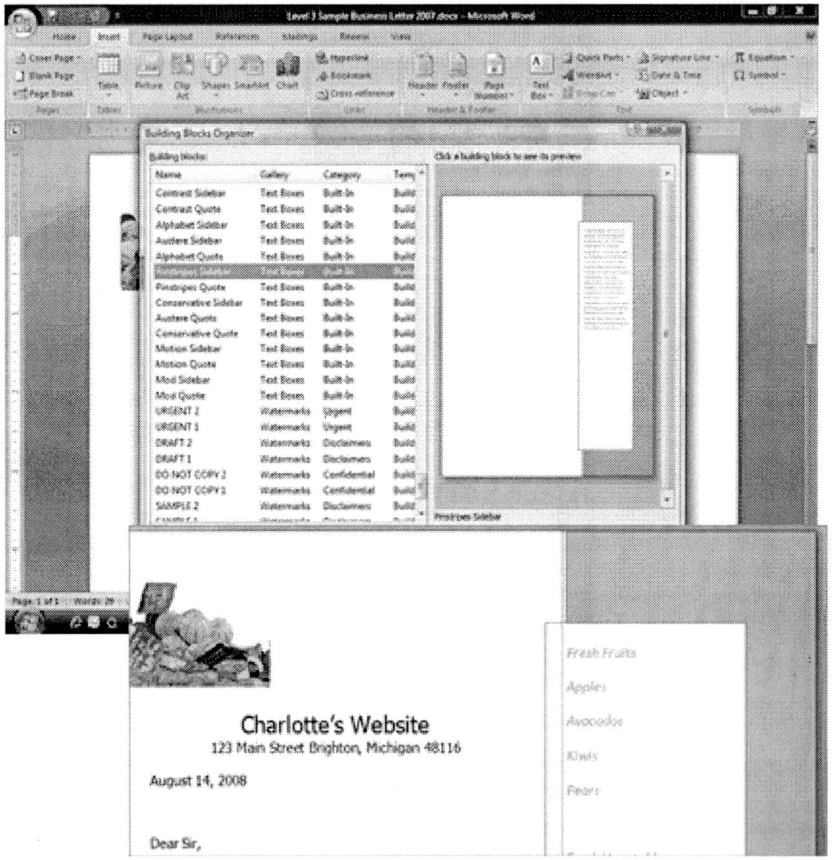

Text Box Tools ->Format ->Styles

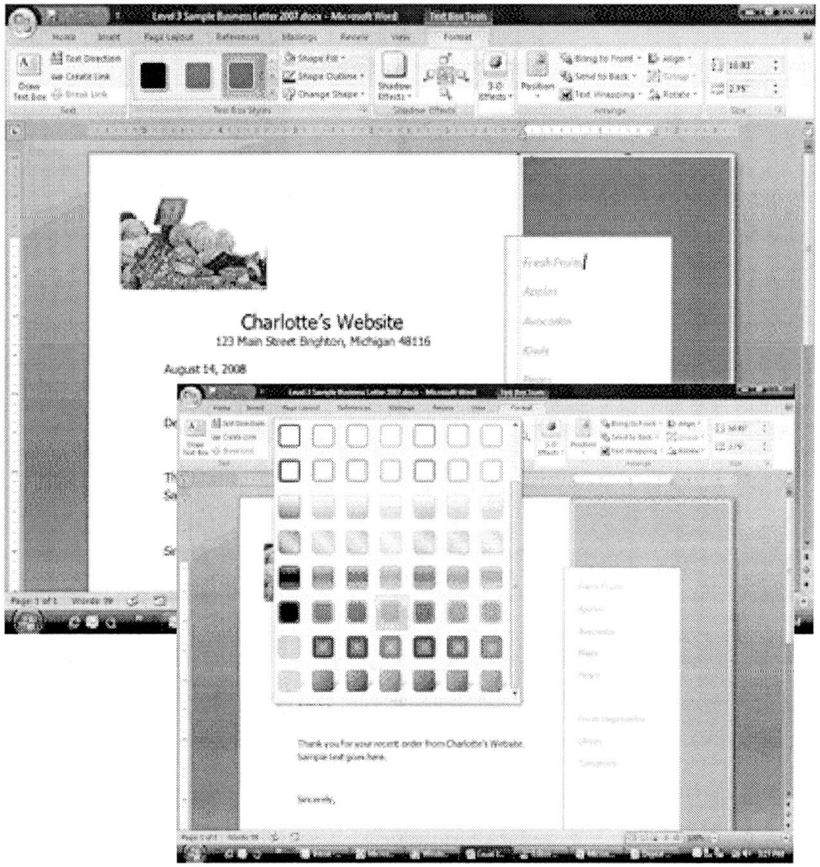

Quick Styles

The new **Side Bar** is a **Text Box** formatted with the Pinstripe Theme. Say you wanted to use seasonal colors on the Side Bar. You can modify the Text Box with **Quick Styles**.

Before You Begin: Place your cursor in the Text Box, please.

Try This. Format the Text Box Style
When you select a Text Box, you should see the **Text Box Tools** at the top of your screen.

Go to **Format ->Styles**.

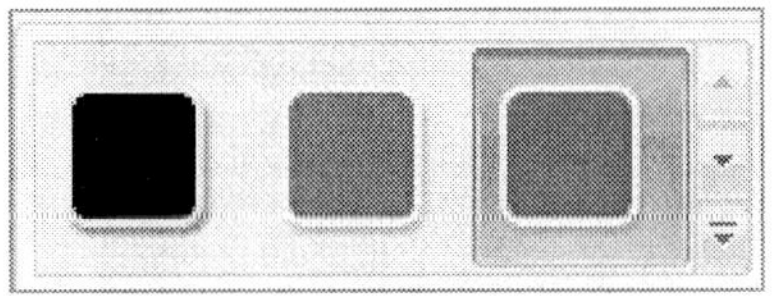

Have Fun...

Modify Building Blocks

Say you wanted to reuse these sidebar themes next season. You can use the **Building Blocks Organizer** to manage different versions of your Building Blocks.

Try This: Modify the Sidebar
Select the Text Box Sidebar.
Go to **Insert ->Quick Parts**.
Select **Add Selection to Quick Parts Gallery**.

Modify the Build Block Properties:
Name: Pinstripes Sidebar.
Gallery: Quick Parts.
You can use the description to document the changes if you wish.

What Do You See? There are two Building Blocks with the same name. However, they are organized into different Galleries. The original Pinstripe Sidebar is a built-in Text Box. The new Pinstripe Sidebar is in the Quick Parts Gallery.

Insert ->Quick Parts -> Add Selection to Quick Parts Gallery

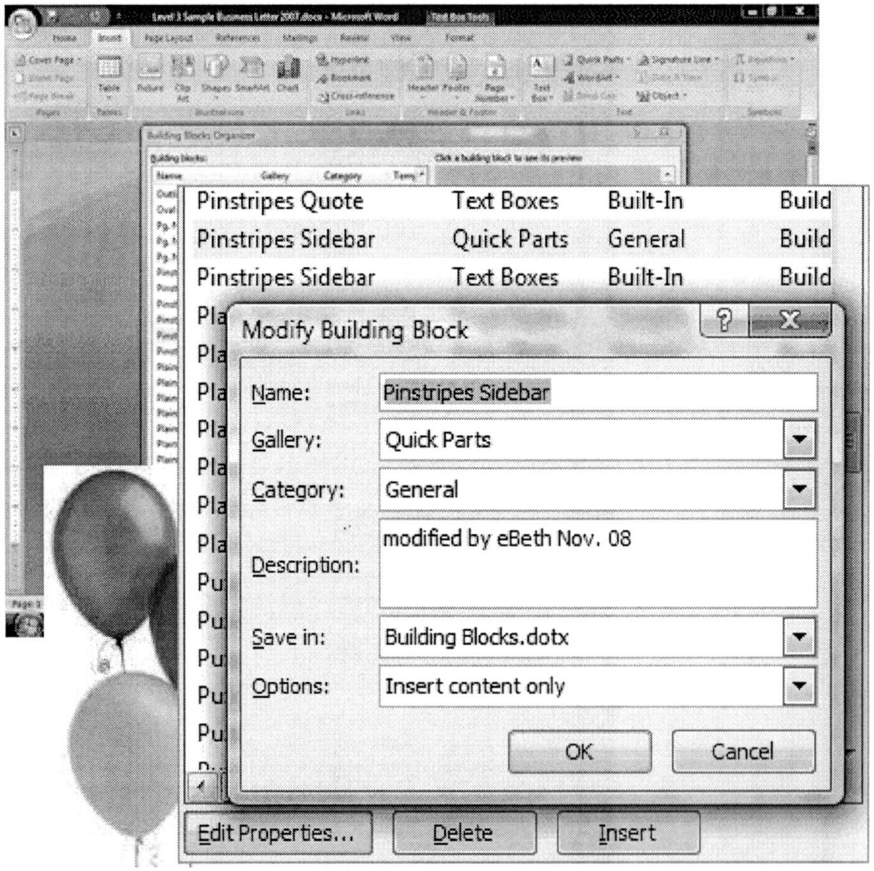

Microsoft Word 2007 Exam 77-601 Topic: 4. Organizing Content
4.1. Structure content by using Quick Parts
4.1.2. Save frequently used data as building blocks: Modify and save building blocks with the same name

Take One

Save Your Work

Please review the main points to keep in mind when you save a file.

1. Where Are You Saving It?
Go to the **Documents** folder.

2. What Is The File Name?
Charlotte's Business Letter 2007.docx

What Do You See?
The new file type in Microsoft Word 2007 is the docx format.

3. What Are You Doing?
Click on **Save.**

Office ->Save

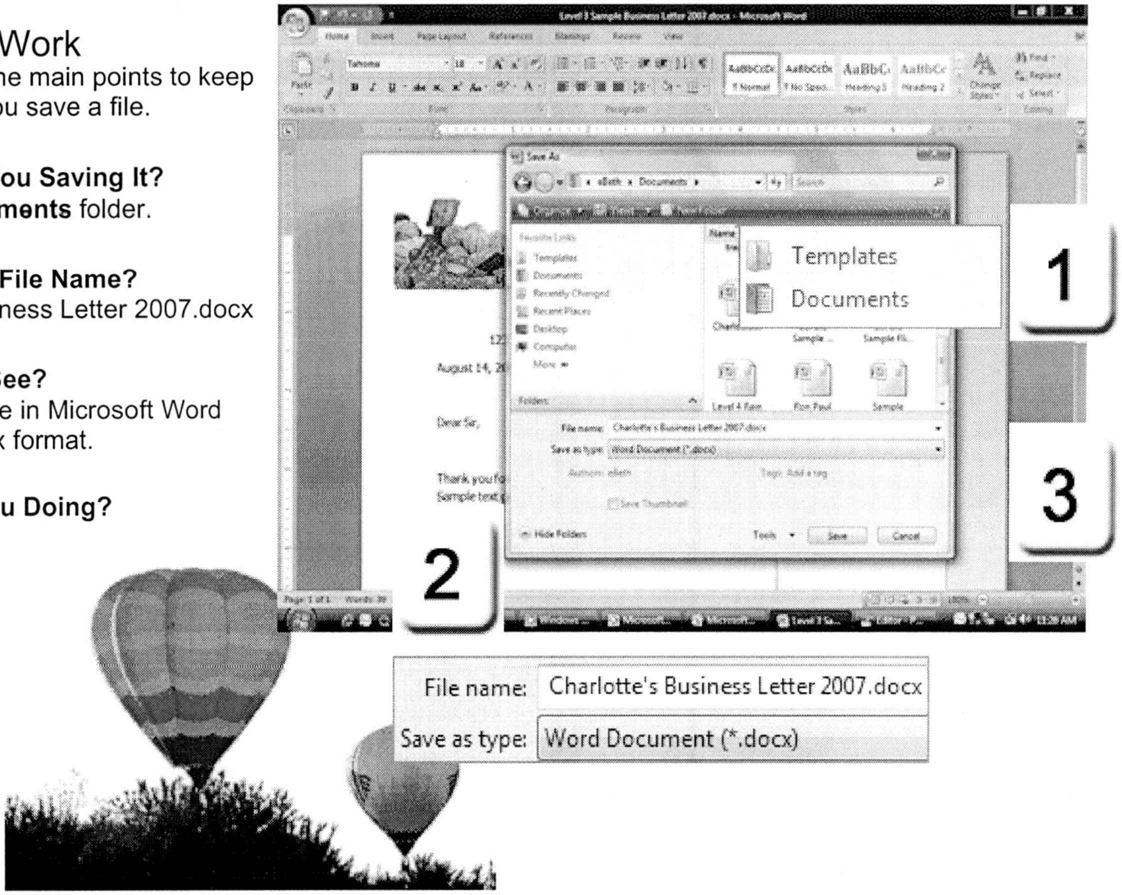

Microsoft Word 2007 Exam 77-601 Topic: 6. Sharing and Securing Content
6.1. Prepare documents for sharing
6.1.1. Save to appropriate formats: Save as a .doc, .docx, .xps, .docm, .or dotx, file

Office ->New ->Blank Document

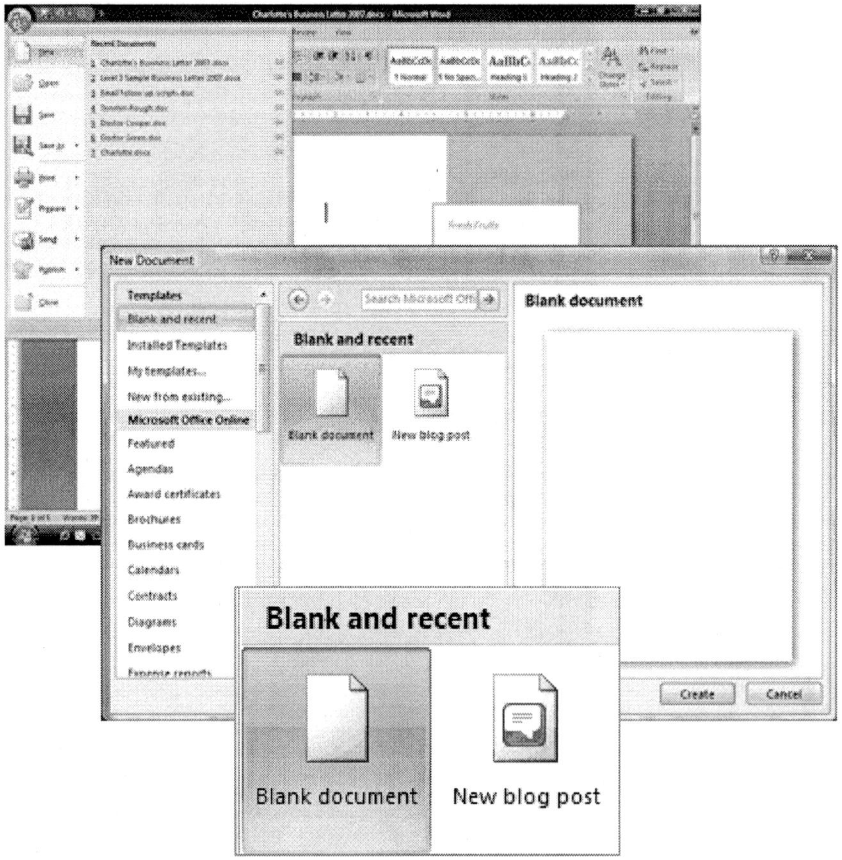

New Document

A **Template** can be a document, spreadsheet or presentation that includes formatting, layout and design. A template gives you a place to start, instead of beginning with a blank page. **Microsoft Office** offers many, many templates that you can use.

Try This: Open A New Document
Go to **Office->New**.

What Do You See? Microsoft Word displays a **Blank Page**. The name, "Blank" is misleading. A plain piece of paper in Word is actually loaded with built in Quick Parts and formatting.

Keep going...

Office ->New ->Installed Templates

New Templates

Try This: Open a Template
Go to **Office ->New**.
Select **Installed Templates**.

What Do You See? The installed templates include professional letters, fax cover sheets, resumes, reports and sample Mail Merge documents.

There are additional templates online including many designs for brochures, calendars and newsletters. Most of the online templates are free.

Good Practice: Open a letter or fax template and edit the Quick Parts. You do not have to save this practice document.

Memo to Self: When you open a new template, you make a copy of the file, just like the Blank Document you open each time you start Word, you need to save your changes.

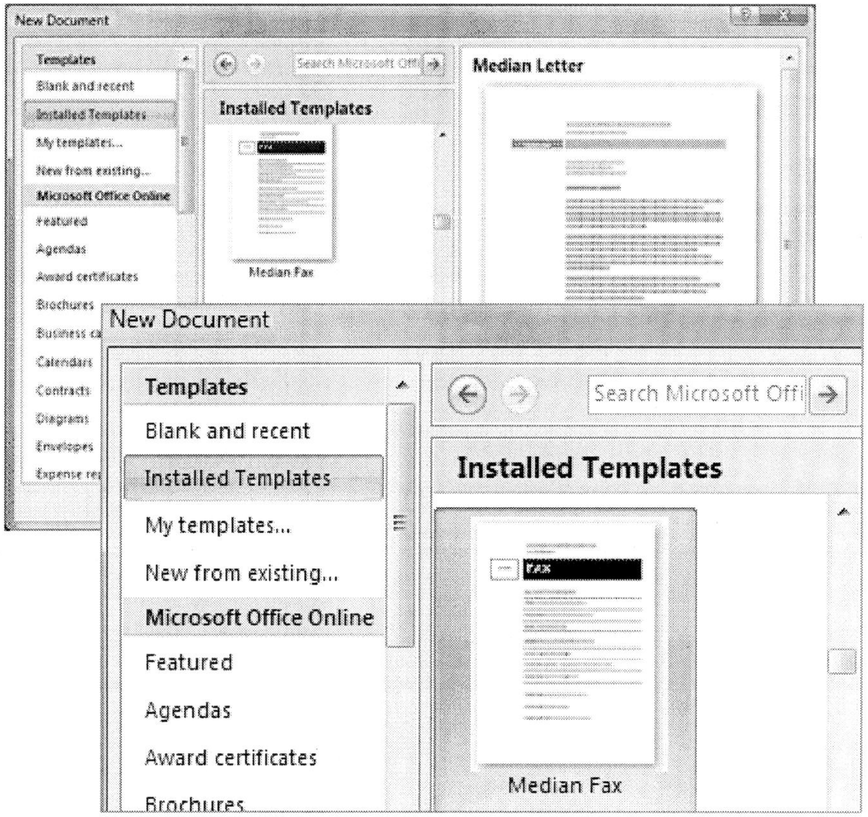

Microsoft Word 2007 Exam 77-601 Topic: 1. Creating and Customizing Documents
1.1. Create and format documents
1.1.1. Work with templates: Create documents from templates

Save As a Template

Say you wanted to use the letterhead you designed for Charlotte's Website to create your own template.

Good, very good. Using your own template saves a lot of time and gives your business or department a professional image.

Before You Begin:
Open the sample business letter you designed previously: **Charlotte's Business Letter 2007.docx.**

Try This: Save As Template
Go to **Office-Save As->Word Template.**

Keep going...

Office ->Save As -> Word Template

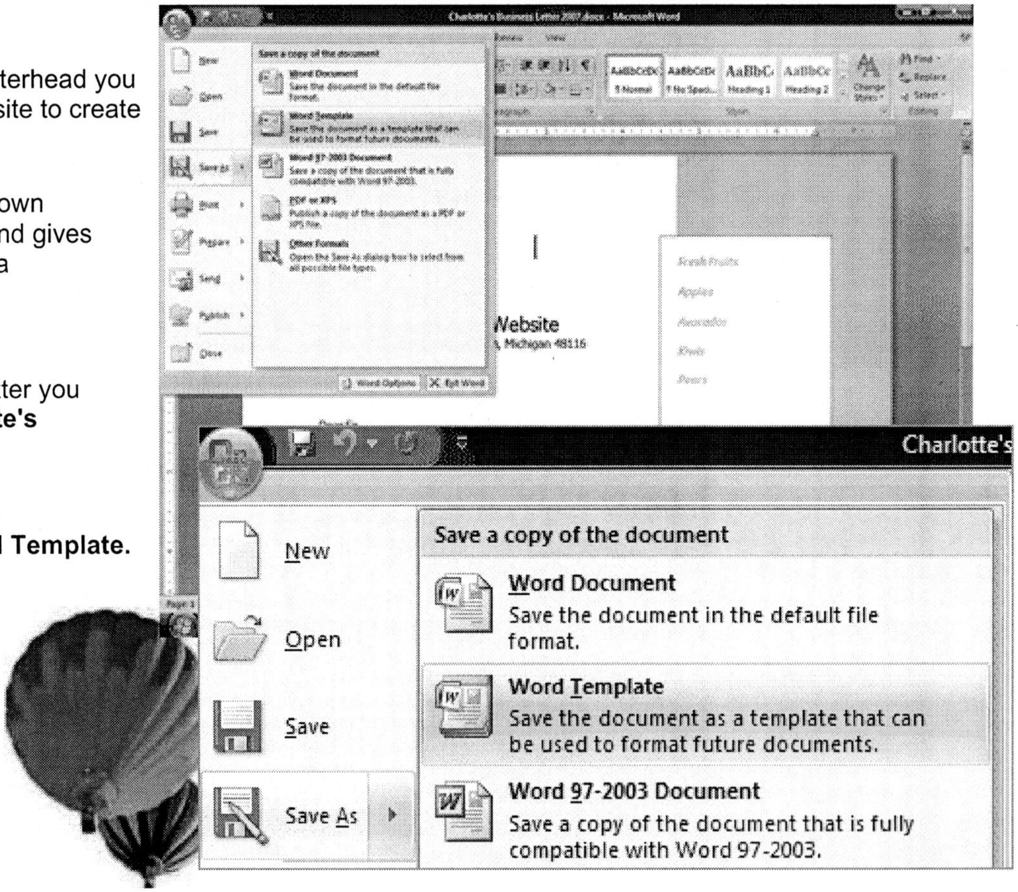

Take One

Save As a Template

Where Are You Saving It:
Select the **Templates** Folder.

What Is The File Name:
Charlotte's Business Letter 2007.

What Do You See?
A Word Template is a *.dotx file type.

Click on **Save**, please.

Keep going...

Office ->Save As -> Word Template

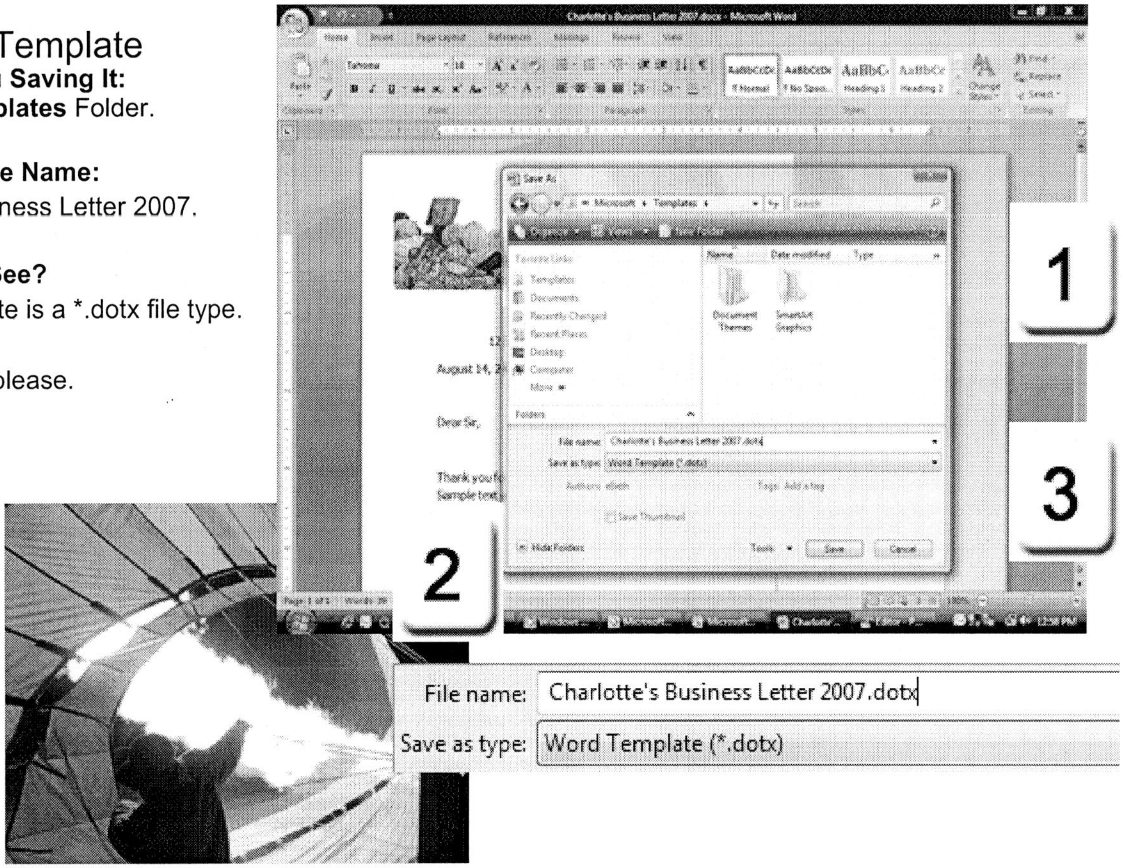

Take One

Use Your Template

So, how do you use your own templates? Here are the steps! ;-)

You Gotta Try This:
Go to **Office ->New**
Select: **Word Template**

What Do You See? The New Document window will open.

What Are Your Choices?
Blank: A white sheet of paper.
Installed Templates: Sample files.
Select **My Templates**.

Go another step...

Office ->New -> Word Template

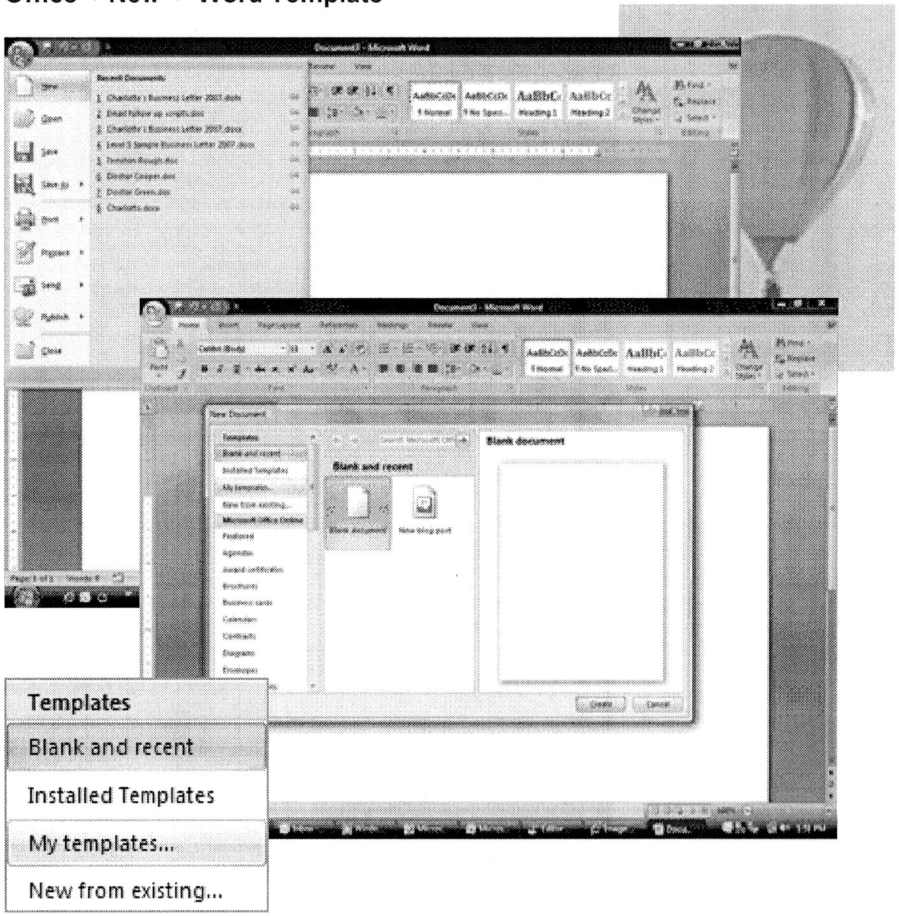

Office ->New -> Word Template -> My Templates

A New Document

When you saved your own business template, you were prompted to select the **Templates** folder as a place to put it. Now, when you go to **Office -> New**, you should see your business letter in My Templates.

Double click Charlotte's Business template and you will open a new document. It should be formatted with a logo and many Quick Parts.

What Do You See? Look at the Title Bar at the top of Microsoft Word. Does it say: Document4? Here is the clue that you are working on a new document which you can save if you wish.

Memo To Self: It may or may not say Document4 depending on how many other files you opened this time. ;-)

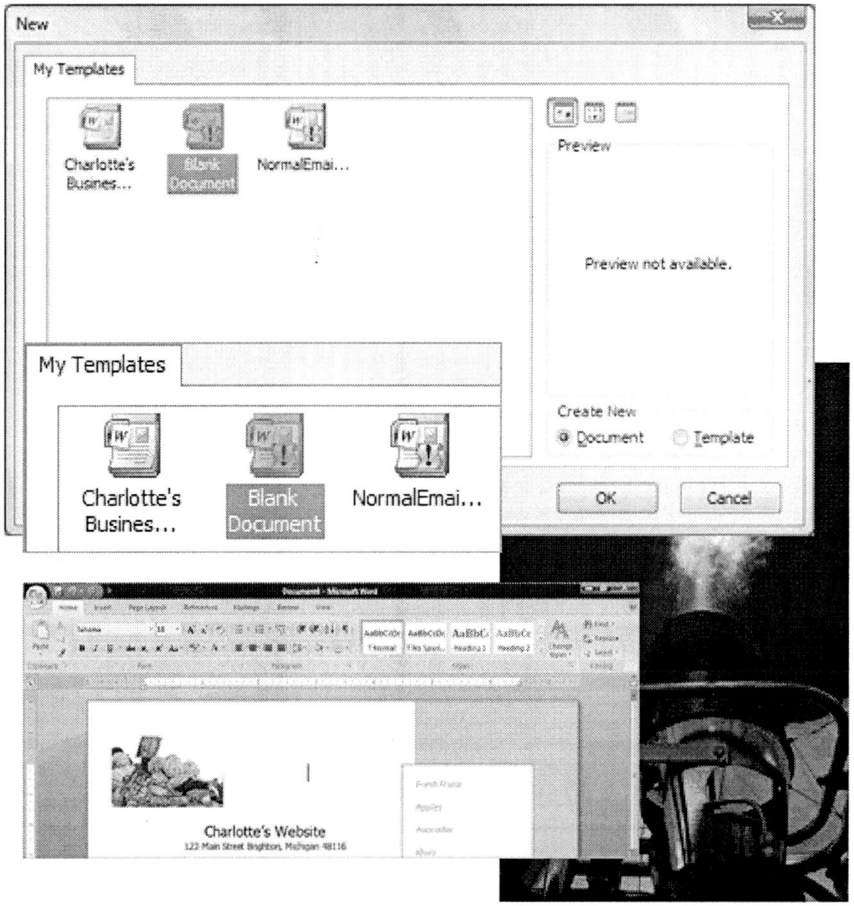

Take One

Test Yourself

Office Efficiency

This chapter began by designing a business letter. We practiced formatting the text and pictures.

In addition, this lesson introduced **Quick Parts** and **Templates**. You can use these options to be more productive and effective with Microsoft Office 2007.

Well...You done good.
You get the cookie.

1. Building Blocks in Microsoft Word are pieces of common text that can be inserted into a document, such as date and time or a greeting line
X a. TRUE
b. FALSE
Tip: Beginning Word, page 81

2. You can create custom building blocks that will appear in the Quick Parts gallery.
X a. TRUE
b. FALSE
Tip: Beginning Word, page 83- 85

3. Building Blocks can only be text.
a. TRUE
X b. FALSE
Tip: Beginning Word, page 83-85

4. What type of file uses the .dotx extension?
X a. A Word Template
b. A Word document
c. A Word table
Tip: Beginning Word, page 86

5. How do you make the Text Box Tools Ribbon appear?
X a. Select a text box
b. Click the little arrow below the Insert Text box button
c. Select the View Text Box Tools command from the View Ribbon
Tip: Beginning Word, page 91

6. Which is true about Templates? (Select all correct answers.)
X a. You can create your own by saving a current document as a template
X b. When you open a template, Word creates a new document based off the template
X c. Word has pre-made templates
Tip: Beginning Word, page 95

Downloads

Graphics: Logo, Biz1, Biz2, Biz3, Flag1, Flag2, Flag3, Fruit1, Fruit2, Fruit3, Peppers1, Peppers2
Completed Business Letter 2007

Assessment

There is no **Skill Test** for this lesson. You can continue to the next lesson if you wish.

Microsoft Word 2007 Exam 77-601 Topic: 1. Creating and Customizing Documents
1.1. Create and format documents
1.1.1. Work with templates

Take Two

WYSIWYG

First Prize

Click Here to Get Started
Sample Files

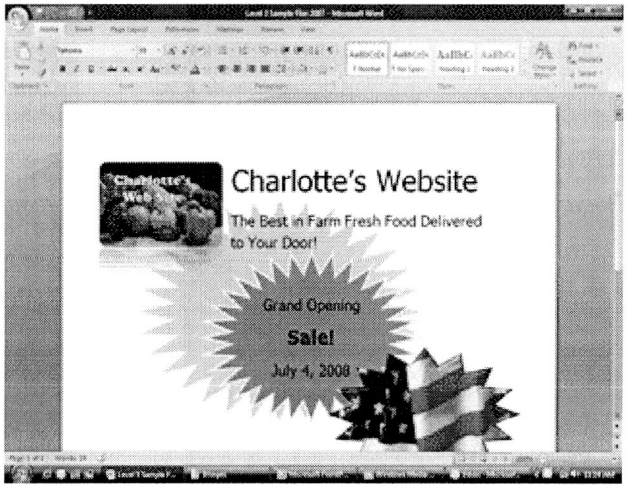

Beginning Word

Lesson Objectives: This demonstration uses a marketing flyer to show additional options for working with visual content: shapes, illustrations and Quick Styles. This example also shows options for reviewing documents. In this lesson you will:

First Prize

Every business and even every department is in a race to win my time and money. Marketing is allowed to "color outside the lines" to get my attention. In addition to setting up formal stationery, you need to promote your products and services. **Start** the **Program** Microsoft **Word**.

What do you see from the top of the screen? Is there a Title Bar that says Microsoft Word? Yes.

Is there a **Home** Ribbon with the **Clipboard, Font and Paragraph** Groups? Yes.

If your screen looks similar to the example on this page, then you are ready to get started.

Home -> Font

Begin with the Text

The five "W's" of good writing are: who, what, when, where, and why. Who are you? What are you saying to me? Why does it matter? All communication and marketing begins with the name.

1. Enter the name

Type: Charlotte's Website

Select the text

Go to **Home ->Font.**

Select: Tahoma, 36 pt

2. Enter the text

Type: The Best in Farm Fresh Food Delivered to Your Door!

Select the text.

Go to **Home -> Font.**

Select: Tahoma, 16 pt

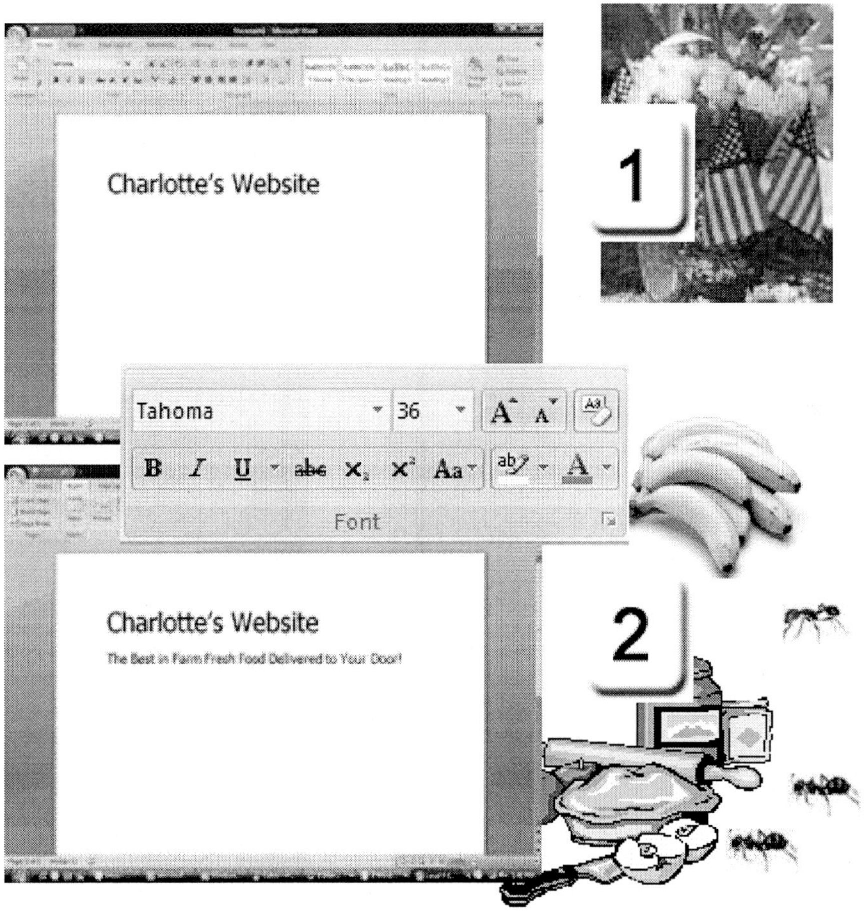

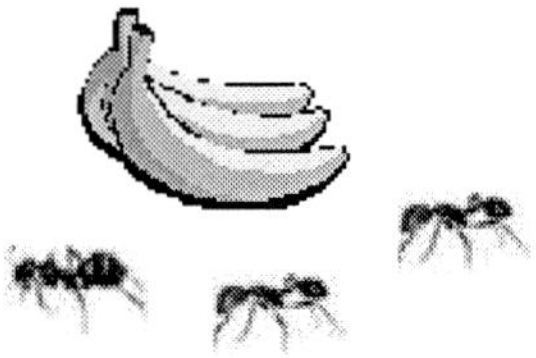

Insert A Picture

There are **sample files** that you can use for this flyer. You are also welcome to use your own imagination.

3. Insert and Format a Picture

Go to **Insert->Picture.**

Look in the **Documents** folder**.**

Select a picture for this flyer.

Go to the **Format** Ribbon.

Choose a **Picture Style** from the gallery.

Change the **Text Wrap** to Tight.

Move and **Resize** the picture.

Insert -> Picture

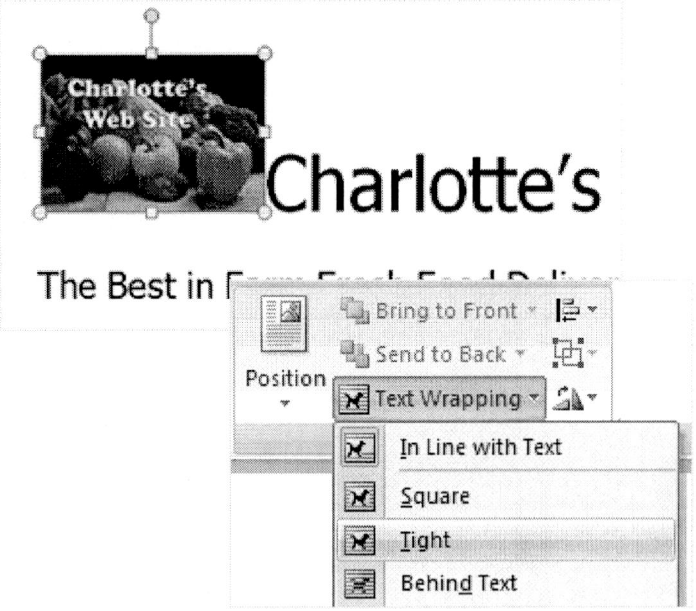

Picture Tools -> Format -> Picture Styles

Picture Styles

Microsoft Word 2007 offers a gallery of **Picture Shapes**, **Borders** and **Effects**. These format options can be found in the **Picture Styles**.

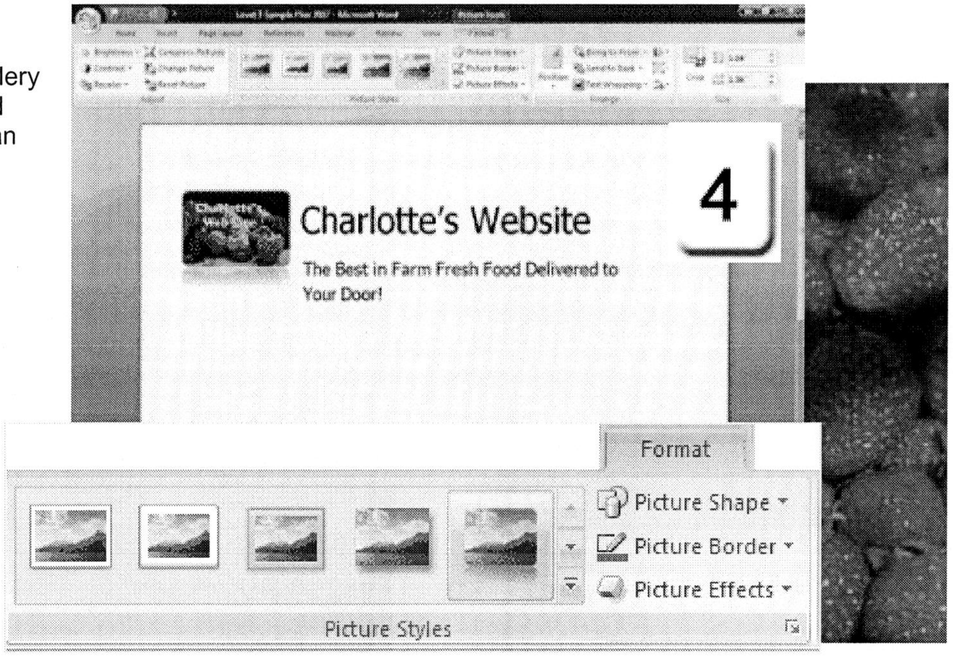

4. Format with Picture Styles

Click on the picture to **Select** it.
Go to the **Format** Ribbon.
Choose a **Picture Style**.

Memo to Self: There are small up and down arrows on the right side of the gallery so that you can see more options.

Microsoft Word 2007 Exam 77-601 Topic: 3. Working with Visual Content
3.2. Format illustrations
3.2.4. Set contrast, brightness, and coloration

Picture Shapes

Picture Shapes takes your pictures and places it inside of a shape. The Picture Shapes and Picture Effects, are part of **Quick Styles**.

The Quick Styles formatting includes the shape, border, shadows and fill effects.

5. Format with Picture Shapes

Click on the picture to **Select** it.

Go to the **Format** Ribbon.

Choose a **Picture Shape**.

Picture Tools -> Format -> Picture Shape

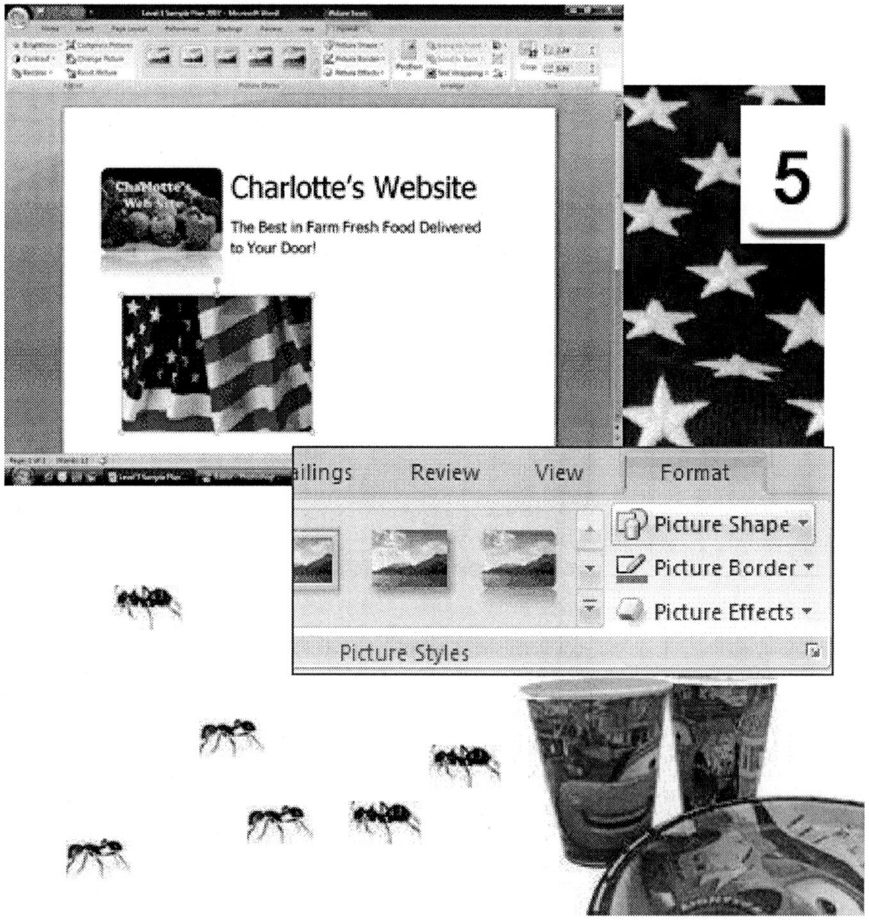

Picture Effects

Picture Effects can be very tasty "eye candy." Please, remember your audience. A little bit goes a long way. Too much candy is simply too sweet.

Picture Tools -> Format -> Picture Effects

6. Format with Picture Effects

Click on the picture to **Select** it.

Go to the **Format** Ribbon.

Choose a **Picture Effect.**

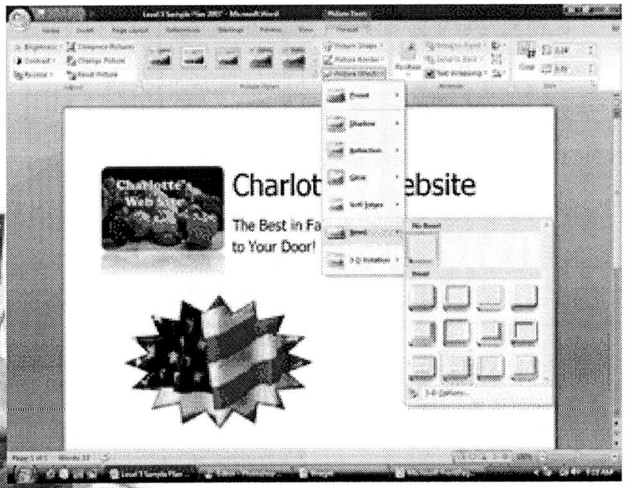

Insert -> Illustrations -> Shapes

Insert Shapes

A **Shape** is a **Textbox** with custom formatting. Microsoft PowerPoint uses Textboxes for the headlines and bullet lists on each slide.

Microsoft Word uses Textboxes to put the Mail Merge fields in the right place for a bar code machine to read the envelope correctly.

1. Insert Illustrations

Go to the **Insert** Ribbon.

Go to the **Shapes** menu.

Choose a Shape.

Hold you cursor on a blank place in your document and **drag a small square**. You should see a shape that you can select, format and resize.

Drawing Tools -> Format -> Shapes

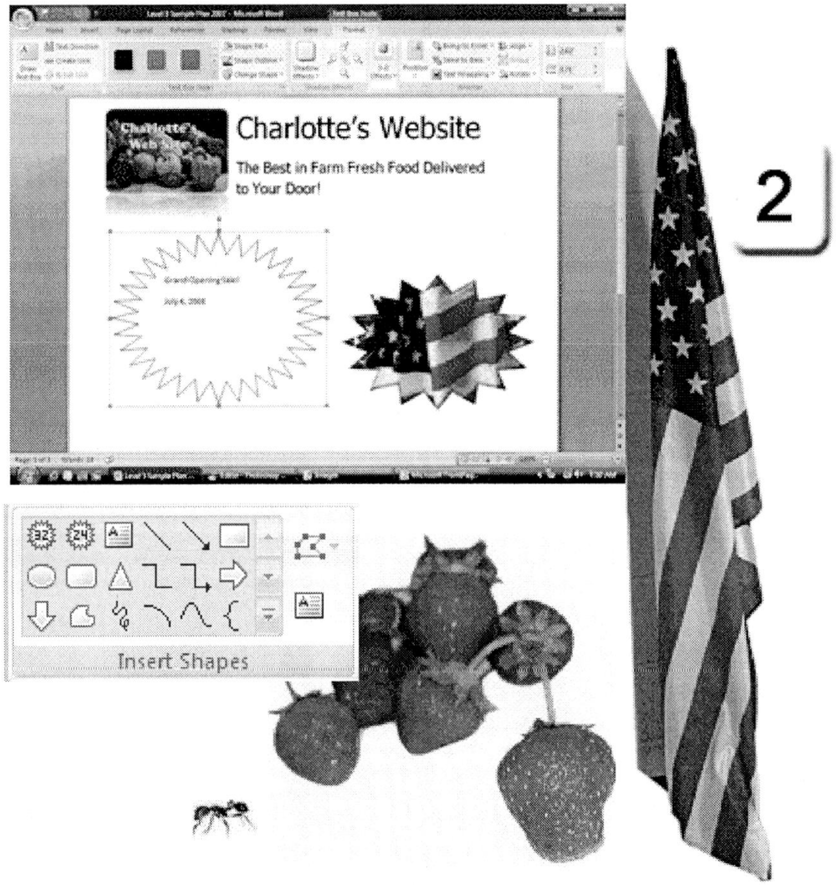

Add Text to the Shape

The Text Box option looks like a flag:

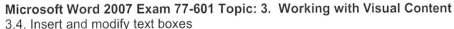

There are two Text Box options. The one in the Group (the blue box of Shapes) creates a NEW Text Box. The one outside of the Group adds text to a shape.

2. Format the Shape (Add Text)
Click on the shape.
You should see the **Drawing Tools**.
Go to the **Format** Ribbon.
Go to the **Shapes** menu.
Choose a Shape.

Hold you cursor on a blank place in your document and drag a small square. You should see a shape that you can select, format and resize.

Text Box Tools -> Format -> Text Box Styles

Format the Text Box

The **Text Box Tools** offers options to **Format** the Fill, Outline and Shadow.

This is the playful side of Word. The purpose of this Text Box formatting is to draw attention to your flyer, and to its message.

3. Format the Text Box

Select the Text Box.

Type: Grand Opening Sale!

Format the type: big, bold, color.

Go to the **Text Box Tools**

Click on the **Format** Ribbon.

Go to the **Text Box Styles.**

Try It: Use the Formatting

Go ahead and use different Fills, Outlines and Shadow Effects.

3

Insert -> Text Box -> Simple Text Box

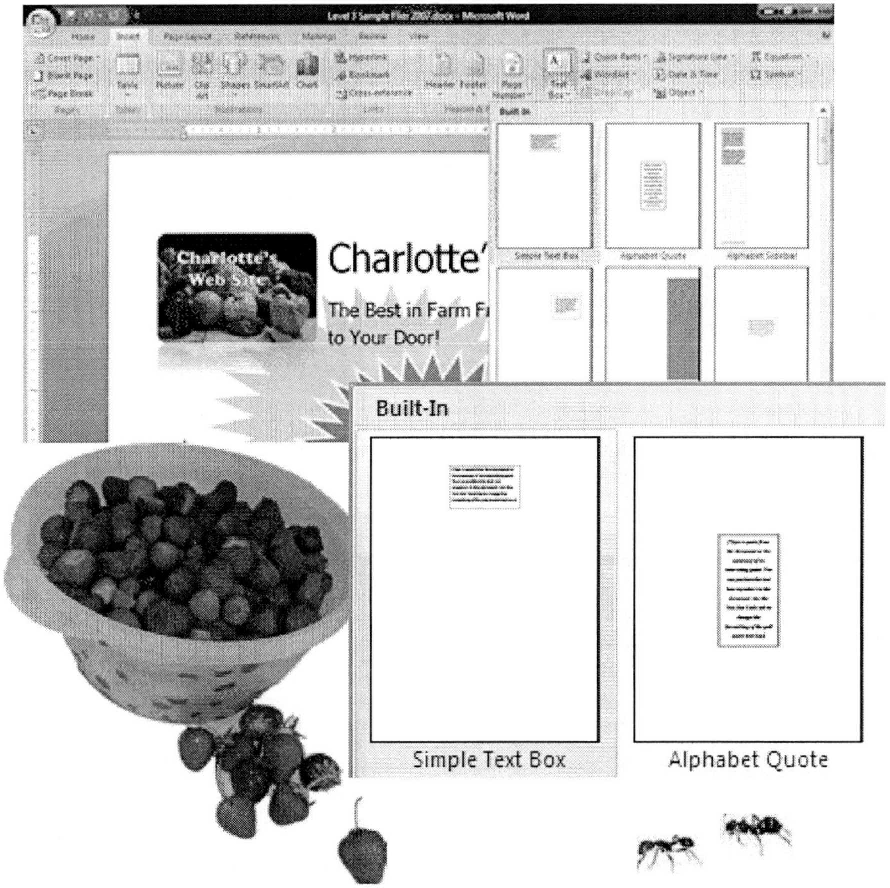

Insert a Text Box

A **Text Box** is a convenient way to add quotes, bling, and other marketing elements to your flyer.

Try This: Insert a Text Box

Go to **Insert ->Text Box.**
Please use the small down arrow to see the **Built-In** boxes.
Select Simple Text Box.

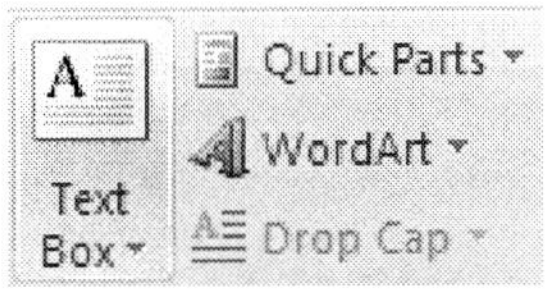

What Do You See? A new Text Box will be placed into your document. Keep going...

Insert -> Text Box -> Simple Text Box

Edit the Text Box

When you click on the Text Box, you should see the **Text Box Tools**. You can move the Text Box if it is covering up your graphics or shapes.

Try This: Edit the Text

Select the Text Box.
Type: You can enter the sentences that you see in this example if you wish.

Try This, Too: Format the Text

Select the word: tomatoes.
Go to Home ->Font.

Can you make just one word big, bold and red and another big, bold and green? Yes. ;-)

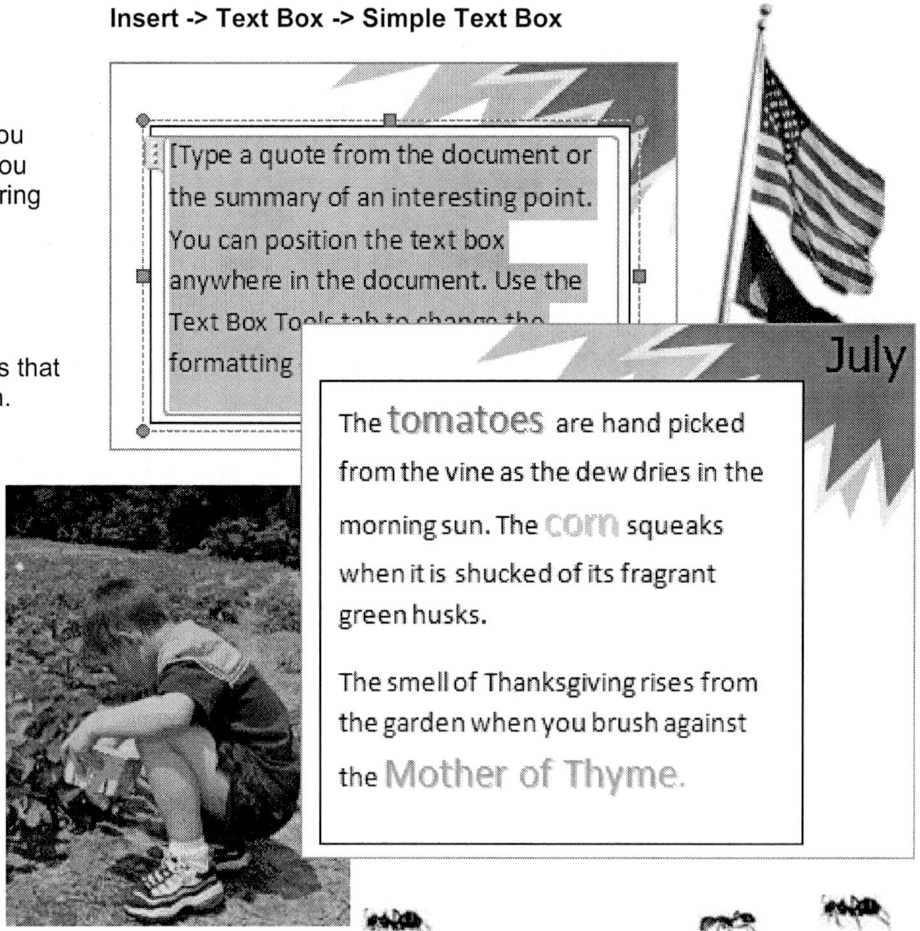

Link Two Text Boxes

You can **Link** two **Text Boxes** together. Say you have a lot of information in the first Text Box. If you link two Text Boxes, then the extra type from the first one will "spill into" the second Text Box.

Try This: Link the Text Boxes
Go to **Insert ->Text Box**. From the options, select **Simple Text Box.**
Select the new Text Box.
Delete the sample type. The new Text Box has to be empty before you can link to it.
Go to **Text Box Tools -> Format**.
Click on: **Create Link**.

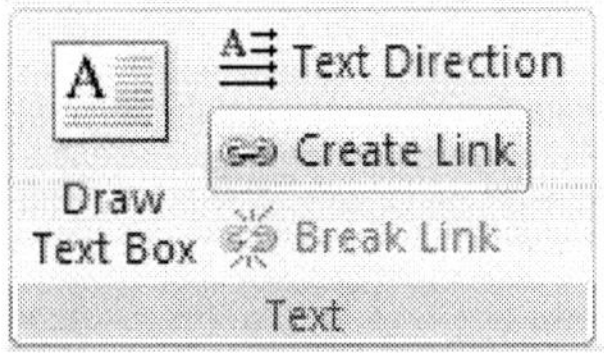

What Do You See? Click on the first Text Box. Your should see a cup that can "pour" the text into the second Text Box. Click on the second Text Box.

Text Box Tools -> Format -> Create Link

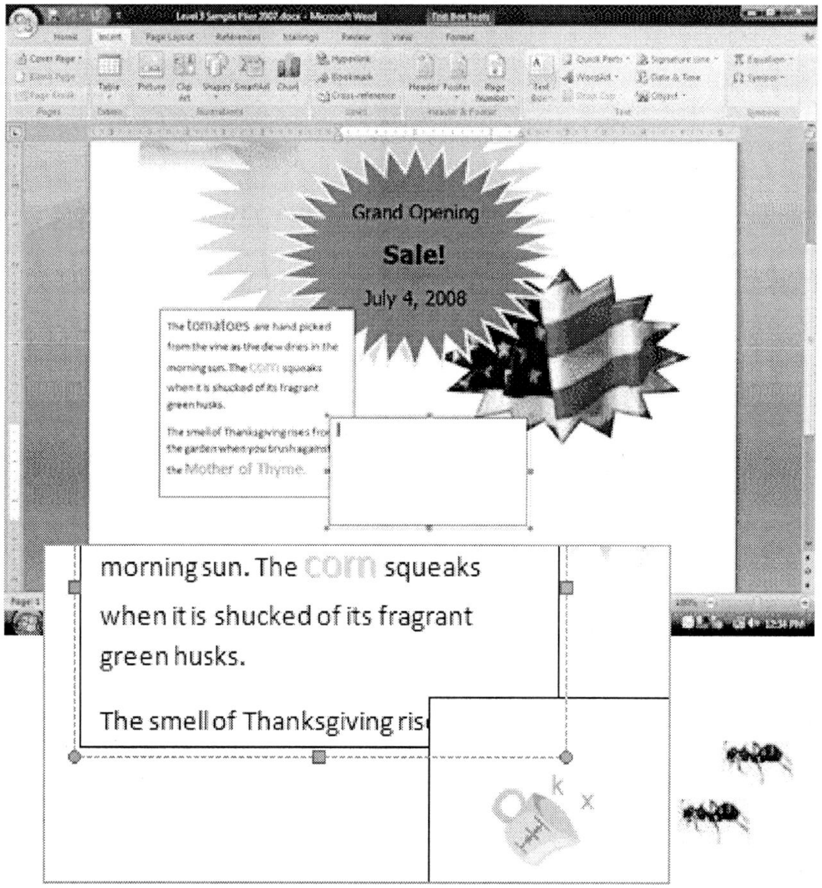

Text Box Tools -> Format -> Create Link

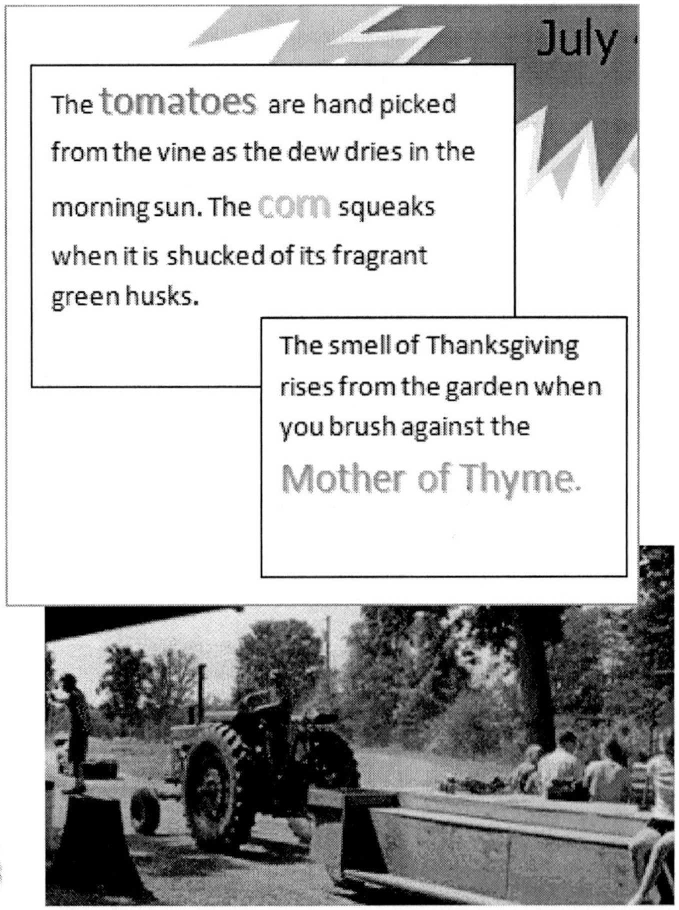

Use the Linked Boxes

Try This: Resize the Linked Text Boxes

Select the first Text Box.
You should see the small handles in each corner and around the sides of the Text Box when you select it.

Resize the first Text Box by using the bottom handles to make it smaller.

What Do You See? As you make the first Text Box smaller, the type shows up in the second Text Box.

Where Have You Seen This Before? Many desktop publishing programs, such as Microsoft Publisher, use linked Text Boxes to create a newsletter. For example, the article may begin on the first page, but the rest of the story may jump (link) to page 5.

Hello SmartArt!

Some **Text Boxes** have additional functionality. In fact, these Text Boxes have so many options that Microsoft named them SmartArt.

SmartArt creates graphics based on your text. An Organization Chart is an example of SmartArt that automatically adjusts the image if you add a new employee.

Before You Begin

Save and close your work.

Go to **Office ->New -> Blank Page**.

Try This: Insert SmartArt

Go to **Insert ->SmartArt**.

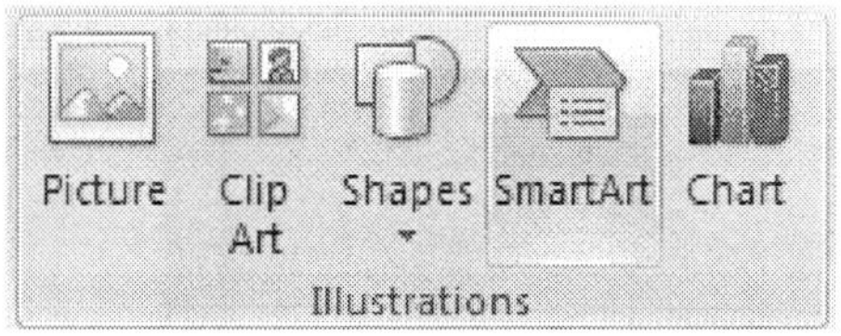

Keep going...

Insert ->SmartArt

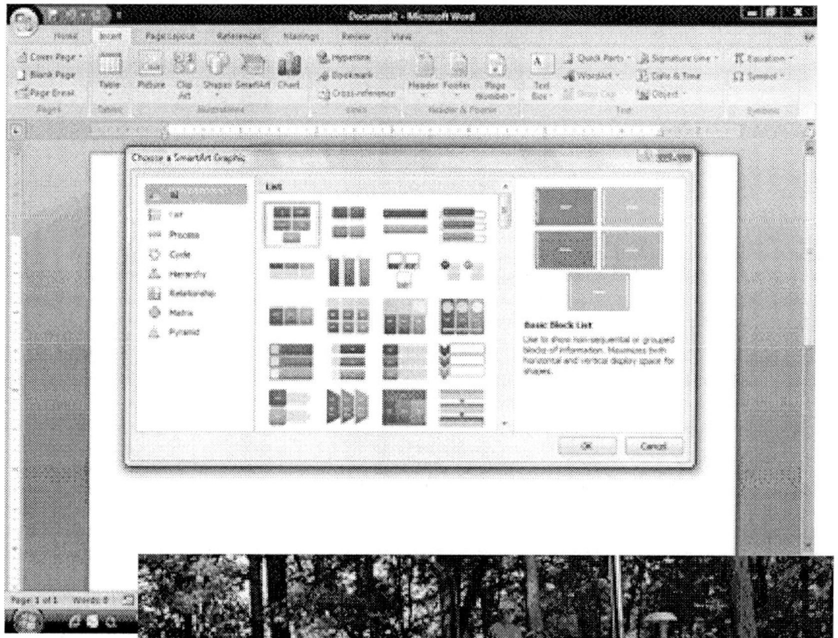

SmartArt Graphics

Choose a SmartArt Graphic

The SmartArt library includes:
List
Process
Cycle
Hierarchy
Relationship
Matrix
Pyramid.

As you click on each option, you should see a sample of the chart, as well as a short description of how you might find this image useful.

Select: Hierarchy.

Keep going...

Insert ->SmartArt

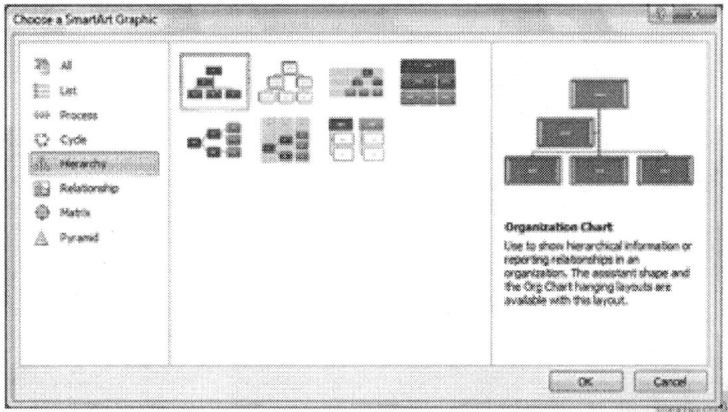

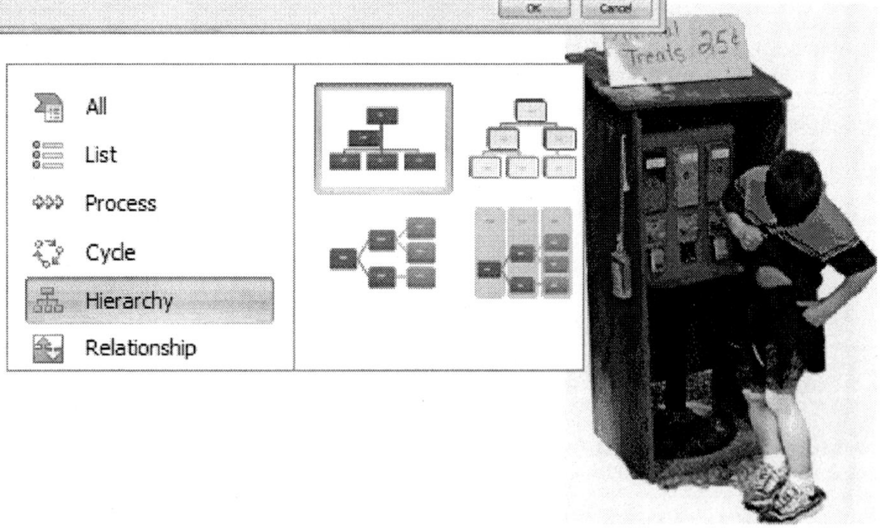

Edit the SmartArt

What Do You See? There are two parts to the SmartArt canvas: the Text Box. and the Diagram. Whatever you type in the Text Box will be displayed in the Diagram.

Try This: Enter Sample Text

Go to the **Text Box.**

Type: Charlotte's Website.

On the second line type: Products.

Please add the following:

Fruit

Veggies

Meat

Cheese

Keep going...

Insert ->SmartArt

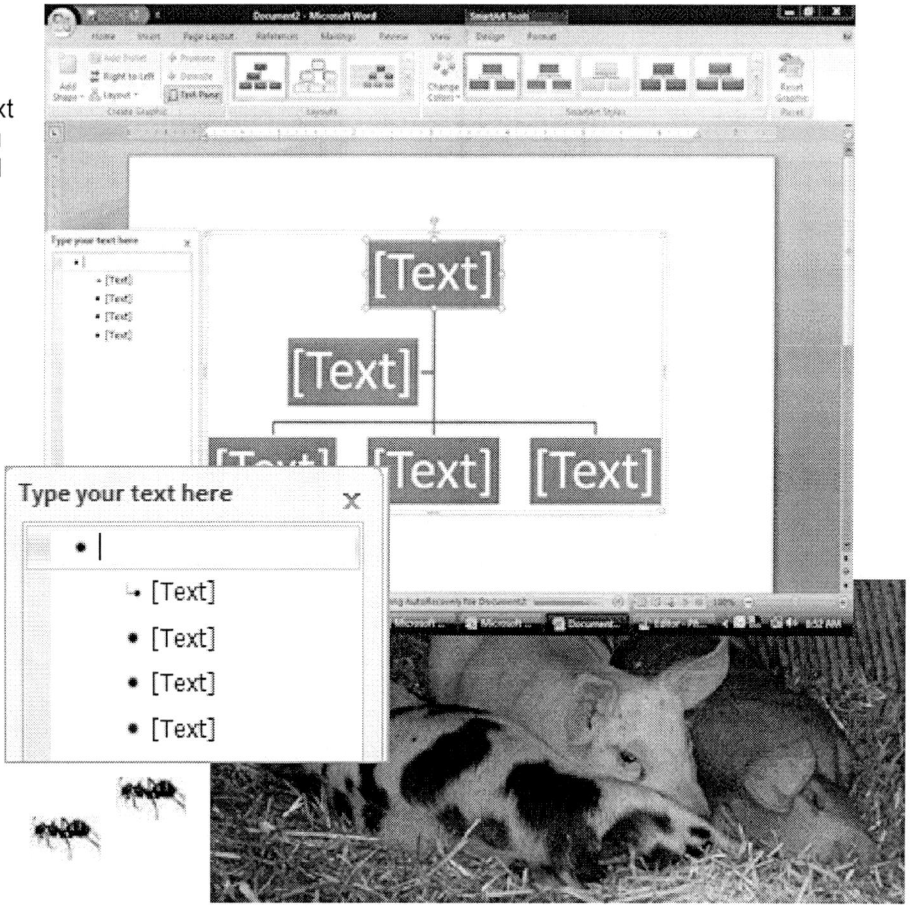

SmartArt Tools ->Design ->SmartArt Styles

SmartArt Tools

The **SmartArt Tools** are similar to the Text Box options. You can use the **Design** Ribbon to edit and improve your chart.

Try It: Use SmartArt Styles

Click on the SmartArt diagram.

Go to **SmartArt Tools ->Design.** Select**: SmartArt Styles.**

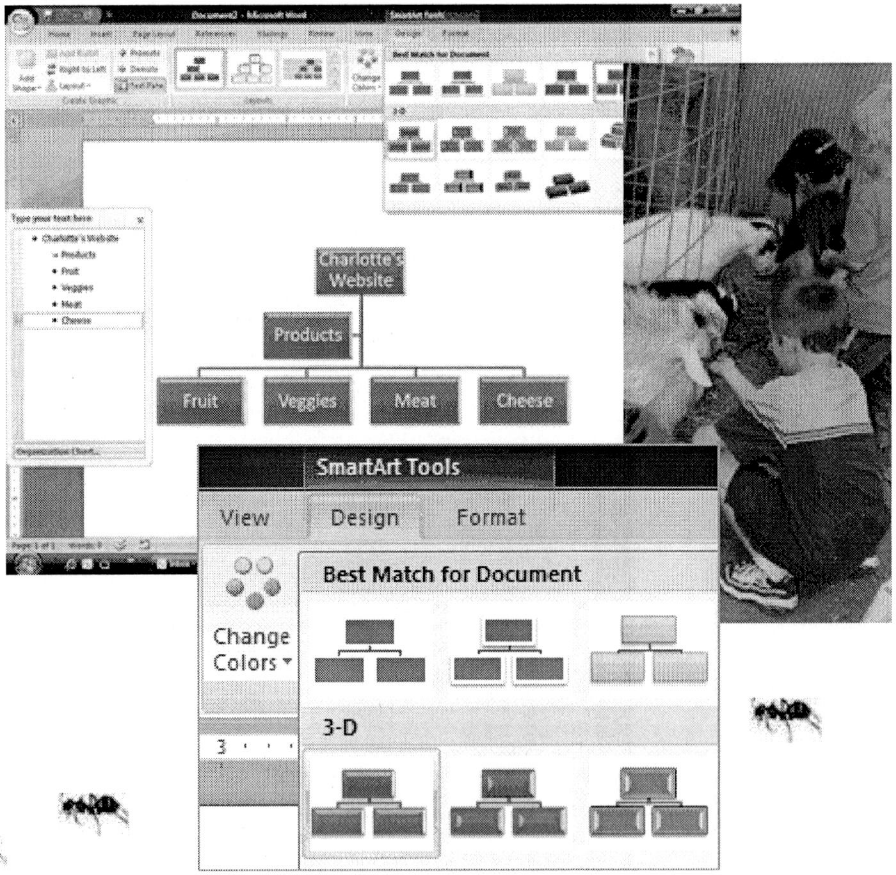

What Do You See? The **Quick Styles** offer a wonderful array of "eye-candy." The Computer Mama enjoys the Polished effect.

Memo To Self: It's called eye-candy because it looks so sweet and sticky, like a tasty confection.

SmartArt Tools -> Design -> Change Colors

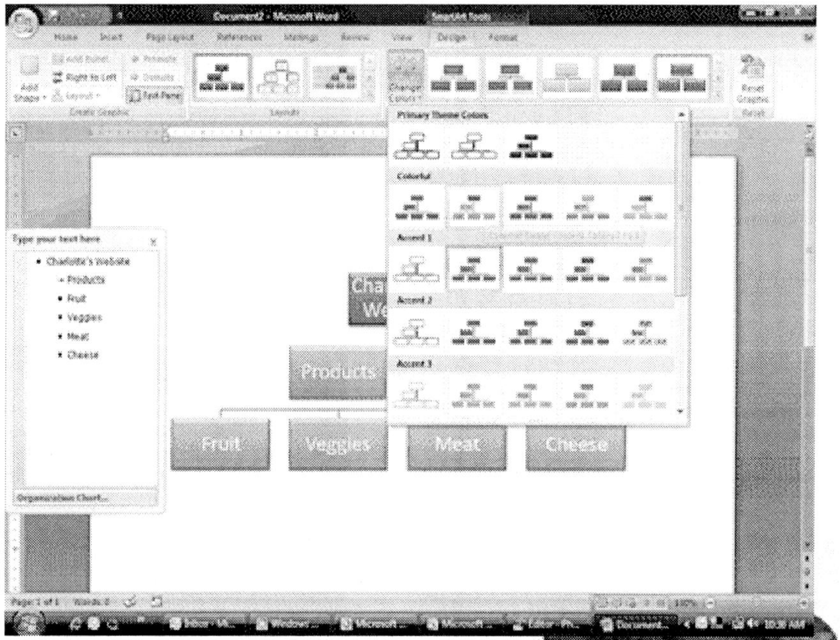

Change the Colors

The **Text Box Tools** sample

Try This: Change Colors

Click on the SmartArt diagram.

Go to **SmartArt Tools ->Dcsign**. Select: **Change Colors**.

What Do You See? The **Style** gallery includes several different formats. As you rub your mouse over any example, Microsoft Word 2007's **Live Preview** will display these colors on your diagram.

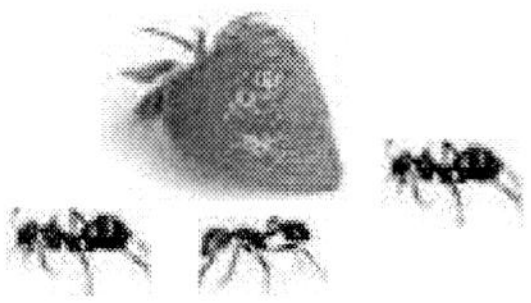

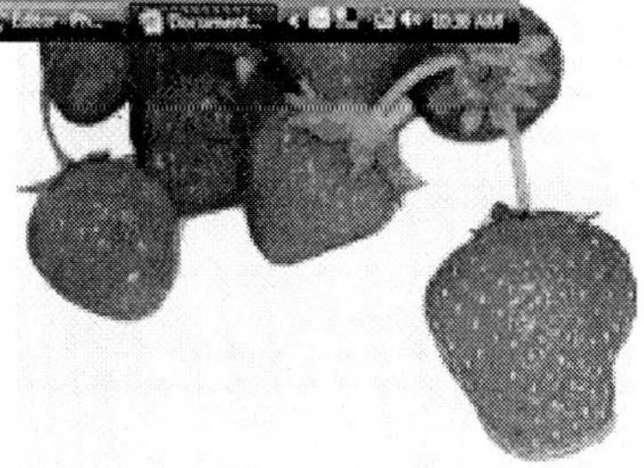

Take Two

Insert ->SmartArt

How Smart Is It?

The **SmartArt** can dynamically adjust to match your text. Say you wanted to add another product to this sample diagram. Here are the steps you would take. Watch what happens. ;-)

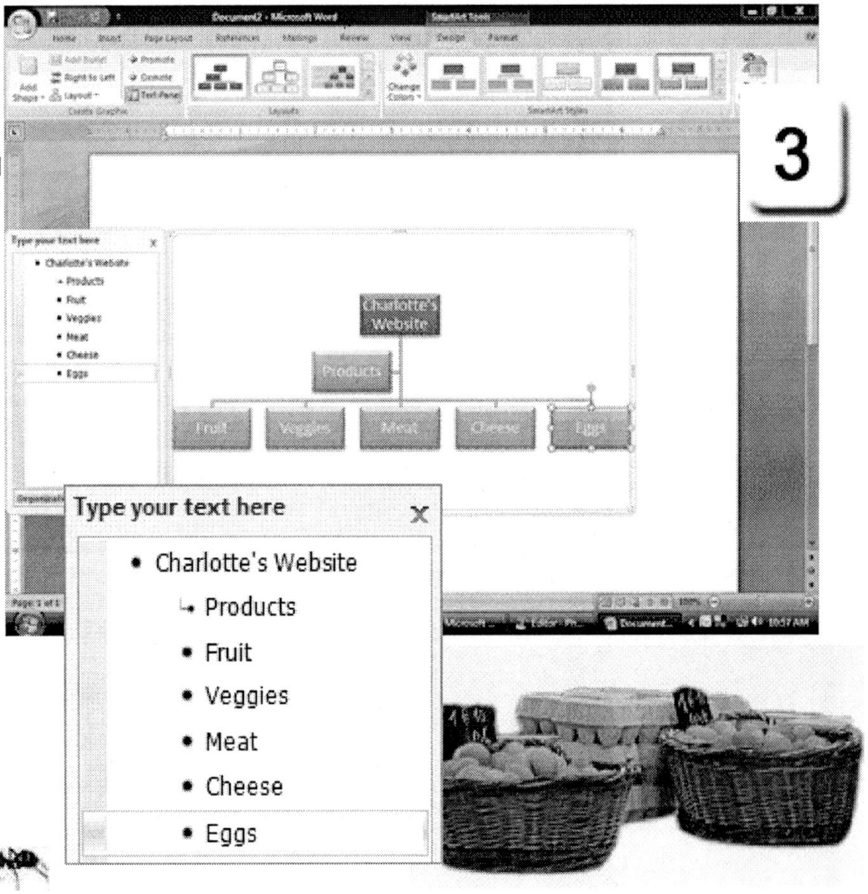

Try This: Edit the Text

Select the SmartArt diagram.

Place your cursor after the word, Cheese and hit the Enter key on your keyboard to create a new, blank line.

Type: Eggs.

What Do You See? The diagram should automatically update to include your new item.

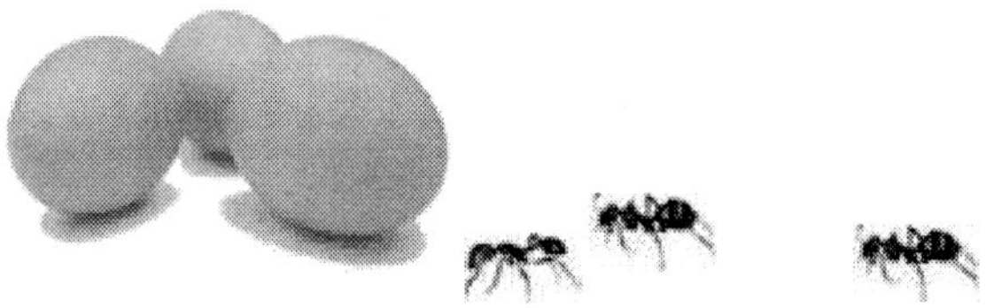

View -> Zoom

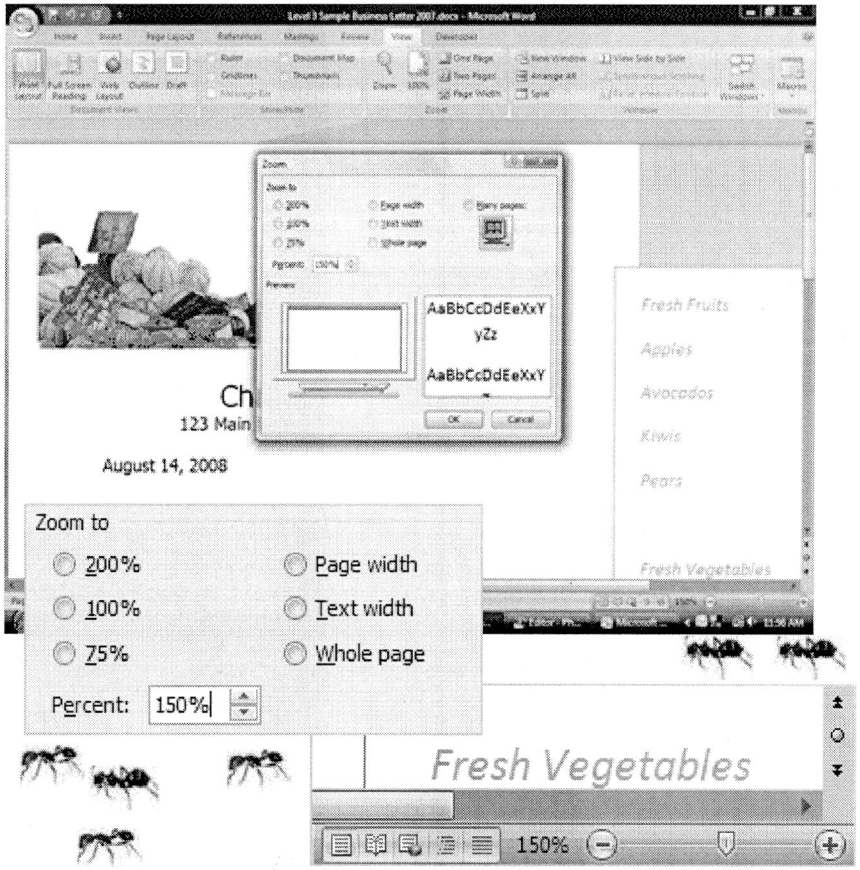

Change the View

There are two ways you can **Zoom** into your work: the **View** Ribbon and the **Zoom Slider**.

The **Zoom Slider** is in the lower right corner of your Word 2007 document. You can drag the slider or use the plus and minus buttons if you wish. The same options are available in the **Zoom** window.

Try This: Change the Zoom
Go to **View ->Zoom**.

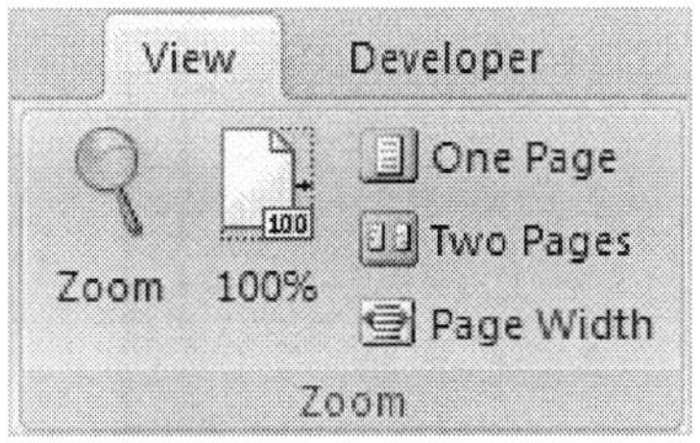

Make it big enough for the old Computer Mama to read, OK?

View -> Window -> Split Screen

Split the Screen

Say you had a long document with many pages that you are working on. You can **Split the Screen** to see two pages at the same time.

Try It: Spilt the Screen

Go to **View ->Zoom.**

Select **Split Screen.**

What Do You See? You should see a line following your cursor. When you click your cursor about half way down the screen, Microsoft Word will display your document in two windows.

Each window can be scrolled and viewed separately.

Remove the Split When you are done editing this mighty document, you can go back to the **View** Ribbon and **Remove the Split**.

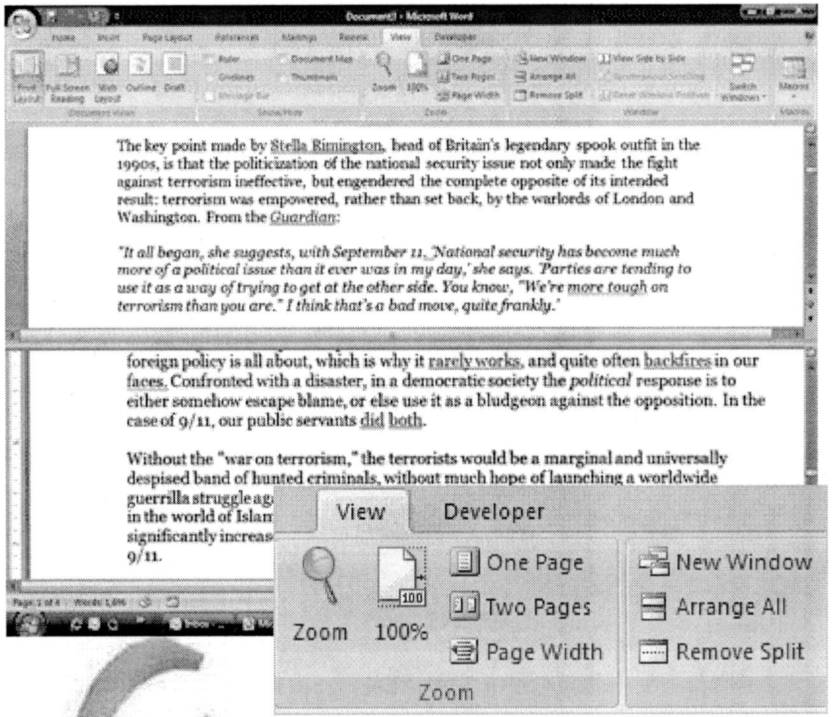

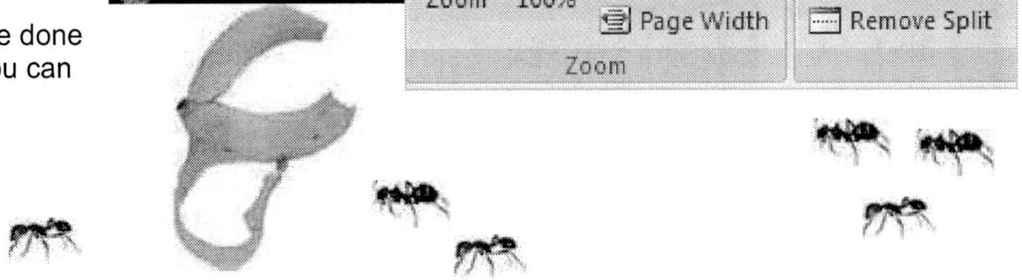

Side By Side

OK, say you wanted to compare two different documents side by side. This example begins by having two sample documents open, please.

Try It: View Side by Side

Go to **View -> Window**.

Select **View Side by Side**.

What Do You See? Two documents that you had opened should be displayed side by side in two different windows. By default, both windows have the same **View** (Zoom) as well as **Synchronized Scrolling**. You can change those options if you wish.

When you are done comparing these files side by side, you can use the **Resizing** buttons in the the upper right corner of Word (the square) to return to a full screen view.

View -> Window -> View Side by Side

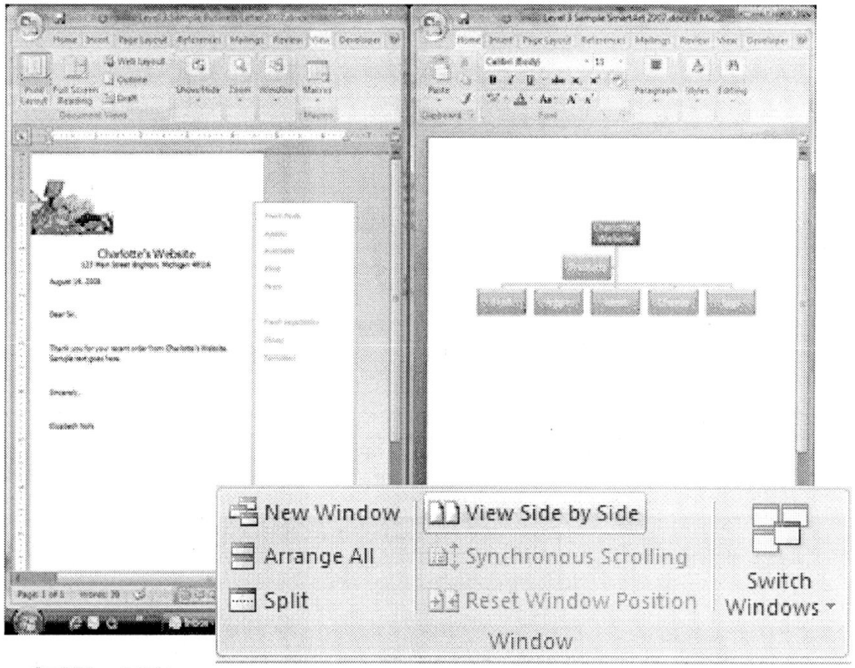

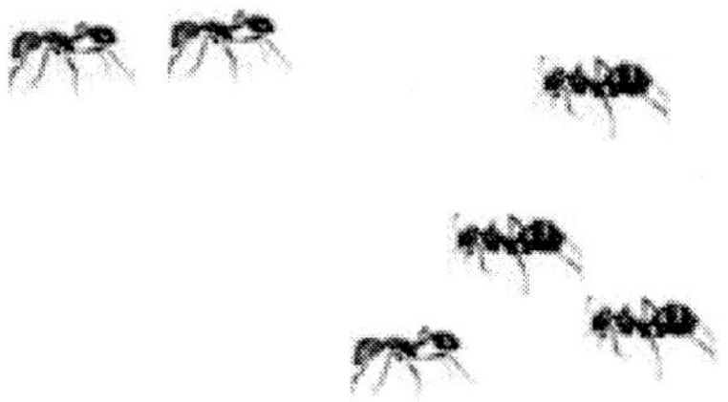

Arrange All

Say you had three (3) documents open at once (multi-tasking!). You can set up Word to display three windows, one on top of another, if you wish.

Try This, Too: Arrange All

Go to **View -> Window.**

Select **Arrange All.**

What Do You See? Three documents will open in three windows. Each document can be scrolled and viewed independently.

Very good.

View -> Window -> Arrange All

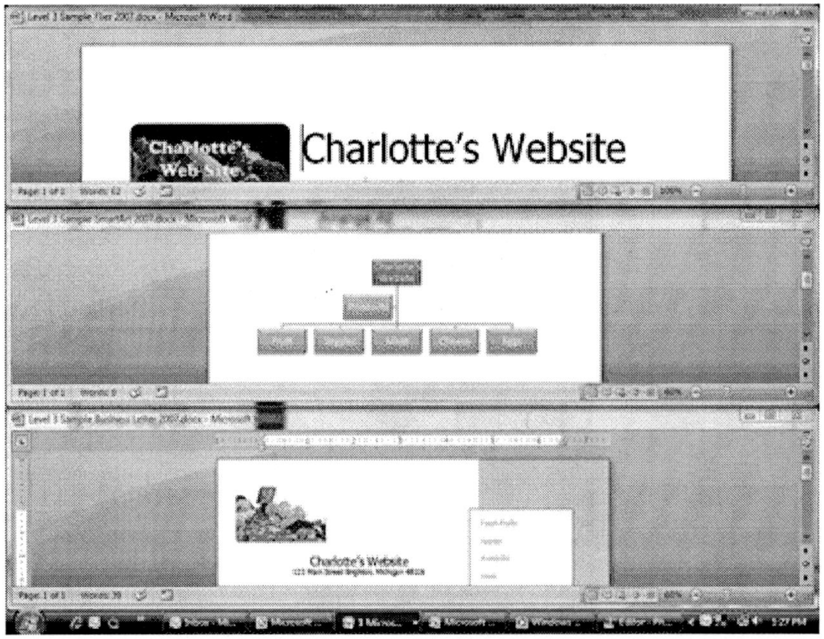

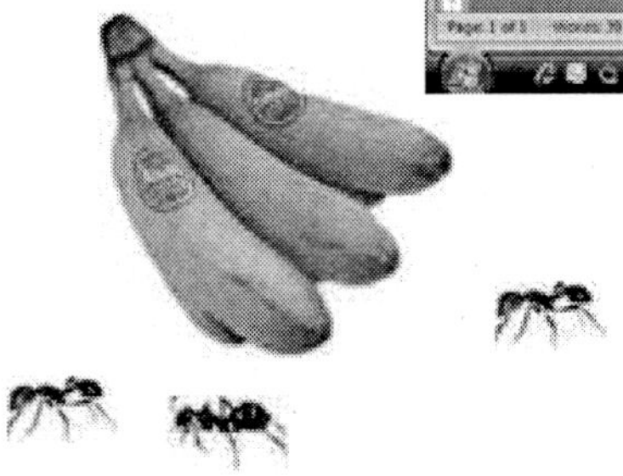

Test Yourself

First Place

Your marketing flyer is competing with hundreds of ads and E-mails. Good products, good company, and a well-done marketing flyer puts you in the race for success.You can save this SmartArt if you wish.

The purpose of this chapter was to learn and practice the formatting options for **Text Boxes** and **Shapes**.

Well, you done good.
You get two cookies.

1. Which are true about Shapes in Word? (Select all correct answers)
a. The Shapes command is on the Insert Ribbon
b. Shapes are text boxes with special formatting
c. You can put text in a Shape
d. You can resize a Shape like a picture
Tip: Beginning Word, page 109

2. If two text boxes are linked, text entered into the first will "spill over" into the second one.
a. TRUE
b. FALSE
Tip: Beginning Word, page 113

3. Live Preview allows you to see changes without applying them to the selected item.
a. TRUE
b. FALSE
Tip: Beginning Word, page 119

4. Which command allows you to compare two documents at the same time?
a. Split Screen
b. Side by Side
c. Reading Mode
Tip: Beginning Word, page 123

5. The Arrange All button will put all open Microsoft Word documents so the windows are one on top of the other.
a. TRUE
b. FALSE
Tip: Beginning Word, page 124

Downloads

Graphics: Logo, Biz1, Biz2, Biz3, Flag1, Flag2, Flag3, Fruit1, Fruit2, Fruit3, Peppers1, Peppers2
Completed Flier 2007

Assessment

There is no Skill Test for this lesson. You can continue to the next lesson if you wish.

Beginning Microsoft® Word: Practice 1

RUBRIC

0	3	5	8	10
Less than 25% of items completed correctly.	More than 25% of items completed correctly	More than 50% of items completed correctly	More than 75% of items completed correctly	All items completed correctly

Each step to complete is considered a single item, even if it is part of a larger string of steps.

Objectives:
The Learner will be able to
1. **Find and use the copy command at least 75% of the time**
2. **Find and use the paste command at least 75% of the time**
3. **Explain that both Word and Excel have the same paste command**
4. **Find and use the cut command at least 75% of the time**
5. **Explain the function of the copy, paste, and cut commands**

Cut, Copy and Paste

Cut, Copy and Paste are basic computer skills. These commands have been part of computers since 1984, long before there was Windows. Each and every program, including Microsoft Word, Excel, PowerPoint and Outlook, uses these functions.

Copy and Paste on the same document

Start the **program** Microsoft **Word**.
Insert a **picture** from **ClipArt**
Select the picture and **copy** and **paste** it five times

Copy and Paste to a different document

Start the **program** Microsoft **Word**.
Insert a **picture** from **ClipArt**
Select the picture and **copy** and **paste**
Go to **New** and open a blank document
Paste the picture into a new blank sheet

Copy and Paste into a different program:

Insert a **picture** in Microsoft Word and **copy** it
Start the **program** Microsoft **Excel**
Paste the picture into Excel

Cut

Cut removes the text or graphic and places it on the clipboard, ready to paste somewhere else. Try it: Insert two new pictures from ClipArt into Microsoft Word. Next to each picture, write what it is. Select ONE picture, go to **Cut**. Open a new blank document and go to **Paste**.

You do NOT have to save these practice files.

Beginning Microsoft® Word: Practice 2

Objectives:
The learner will be able to:
1. **Find and use the Insert Clip Art command at least 75% of the time**
2. **Select and resize a picture using the picture handles**
3. **Use the alignment buttons to center and left align text and pictures**
4. **Insert a Date and Time Text field from the Insert Menu**

Create a Business Letter

A "Corporate Stripe" is a set of documents that have the company logo, fonts, and styles. This exercise allows you to practice formatting text and pictures while you create a business letter.

Type the company name and address
Open a blank Microsoft Word document. Type the following information:

Computers Are Us
555 Main Street
Brighton, MI 48116
(810) 555-1212

Select All of the text and use the Font options to format the type:

Tahoma, 12 point, bold, centered, and dark red

Select the first line of type and make it 14 point.

Insert a Picture from ClipArt
Go to Clips Online, the Microsoft Design Gallery

Search for a photo or cartoon of a computer
Select two or three images and Download them

Use one of the pictures for a company logo
Resize the picture
Center it above the Company name and address

Insert the Date and Time
Remember, the default Date and Time updates automatically. This option is not appropriate for medical or legal documents that must be date/time stamped, but is fine for this exercise.

Type a sample business letter:
Dear Sir,

Thank you for your order, yesterday. We will be shipping your parts by FedX ground.

Sincerely,

Your Name

Save your practice document and name it: Beginning Word Practice 2

Beginning Microsoft® Word: Practice 3

Objectives:
The Learner will be able to:
1. **Resize a picture to exact measurements using the Format Picture command**
2. **Change the text wrapping**
3. **Insert pictures with Insert Clip Art command**
4. **Apply a border to a picture**
5. **Crop a picture using the Format Picture Command**

Working with Pictures

Open a blank Microsoft Word document. You can use Microsoft ClipArt, or Clips Online, to do the following practice exercises.

Insert a picture of a sun or sunset.
Use **Format->Size** to resize the picture to 1.5" wide
Use In-Line **Text Wrapping**
Next to the picture **type**: The weather is great!

Insert a picture of a camera.
Change the Text Wrapping to Tight
Resize the picture to be 2.5 inches tall
Place the picture to the bottom of the page

Insert a picture of a beach.
Format Text Wrapping to Tight
Place the picture into the center of the page
Add a thick BLUE border around the picture
Crop the picture .5 inches from the left

Save your practice document and name it: Beginning Word Practice 3

Beginning Microsoft® Word: Practice 4

Objectives:

The Learner will be able to:

1. **Format text color, bold, and size at least 75% of the time**
2. **Insert a file INTO an existing Word document**
3. **Format text into columns**
4. **Insert a picture from Clip Art and the Design Gallery Live at least 75% of the time**
5. **Change text wrapping around a picture at least 75% of the time**
6. **Apply borders and shading to a whole page using the Format Borders and Shading command**

Create a Flier

Make the headline for the flyer

Type the words: Cub Scouts Enjoy Fall Hike

Format the headline big, bold, centered and dark red

Enter two blank lines after the headline

Type: Sample text

Does the Formatting stay big, bold, centered and dark red?

Change the formatting to Arial, 11 pt, aligned left, black

Download the sample text file from the online course

Save the Cub Scout Text file in your Documents folder

Go back to the Cub Scout flier in Microsoft Word

Go to Insert ->Text Object ->Text from File

Look in your Documents folder for the Cub Scout Text file

Double click the Cub Scout Text file to insert it into the flier

Create two columns of text

Select all of the text EXCEPT the headline

Go to Page Layout -> Columns

Select the option for 2 columns

Add pictures

Insert a Picture from ClipArt

Look for pictures of leaves, scouts and fall

You can use the Design Gallery Live for photos

For each picture, change the text wrapping to be tight

Format Borders and Shading

Before you add borders and shading, notice what you have selected on your document. If the picture is selected, then you will be adding borders to it. If you have text selected, then the border will be added to your word(s). To add Borders and Shading to an entire page, make certain you have NOTHING selected!

The default is NONE. Select a Box, Shadow, etc from the right hand side.

To change line style, choose a format from the middle Style window.

To have little pictures instead of a line, select Art and browse through the options.

Width indicates how thick the line (or art) can be.

Click OK to complete this action.

Save your practice document and name it: Beginning Word Practice 4

Name: ______________

Instructor: ______________

Test Score: ______________

Microsoft Skills Test

Beginning Word

☐ 1. Microsoft Word: Action Step 1
Start the program Microsoft Word.

☐ 2. Microsoft Word: Action Step 2
Type your First and Last Name

☐ 3. Microsoft Word: Action Step 3
Select your First Name and make it big (14 pt), bold, and blue.

☐ 4. Microsoft Word: Action Step 4
Select your Last Name and delete it

☐ 5. Microsoft Word: Action Step 5
Change your mind and Undo the deleted Last Name.

☐ 6. Microsoft Word: Action Step 6
Type the following sentence, including the misspelled word:
We was walking to schoool.
Correct the spelling and grammar errors using the Microsoft tools.

☐ 7. Microsoft Word: Action Step 7
Add two blanks lines at the top of the document. Place your cursor on the top line and insert today's date.

☐ 8. Microsoft Word: Action Step 8
Go to the end of the document and insert a business picture from ClipArt.
Change the text wrap to Tight and resize the picture to be about 1" x 1".
Move the picture to the upper left corner like a logo.

☐ 9. Microsoft Word: Action Step 9
Save the file as Your Name Word Beginning Sample.
Please submit the Sample document to your instructor.

Name: ____________

Instructor: ____________

Test Score: ____________

Microsoft Skills Test

Beginning Word

1. Beginning Microsoft Word
Which of the following are true?
Indicate all the correct statements with a check mark in the box

- [x] A. To open a new Word document, go to File on the menu bar then New
- [ ] B. To open a new Word document, go to File on the menu bar then Open

2. In order to format text in Microsoft Word, you have to select the text you want to change. Check all that are true

- [x] A. Double Click selects a word
- [x] B. Triple click selects a paragraph
- [ ] C. Single click selects a letter

3. Toolbars: which of the following are true?
Indicate all the correct statements with a check mark in the box

?
- [ ] A. All toolbars can be turned on or off with the View -> Options menu command
- [ ] B. All toolbars can be turned on or off with the View -> Toolbars menu command

2003 only Question

4. Spelling and Grammar: A red wavy line under a word means that it is does not match a word in the dictionary.

- [x] A. True
- [ ] B. False

5. Spelling and Grammar:
A green wavy line indicates a grammar check.

- [x] A. True
- [ ] B. False

6. Spelling and Grammar:
If a word is not in the dictionary you can add it by right clicking the red wavy underline and selecting Add from the options list.

- [x] A. True
- [ ] B. False

7. Working with Graphics:
To add a picture that you downloaded from your digital camera to your hard drive, the command is File -> Open.

- [ ] A. True
- [x] B. False

2007 press on insert picture and it takes you to your files.

Microsoft Assessment Test

Beginning Word

8. Working with Graphics:
You resize a picture by dragging any one of the handles

- [x] A. True
- [ ] B. False

9. Working with Graphics:
To make a watermark, select the picture. Go to Format Object, and select the Picture Tab.

- [ ] A. True
- [x] B. False

10. Beginning Word:
In order to format text in Microsoft Word, you have to select the text you want to change: check all that are true:

- [x] A. The backspace key on the keyboard deletes the letters to the left one at a time
- [x] B. The delete key on the keyboard deletes the letters to the right one at a time
- [ ] C. The cursor in Word is an I-beam that follows your mouse
- [x] D. You can use the arrow keys on the keyboard to get to the place you want in a word or sentence.

11. Working with ClipArt:
Microsoft Office has a ClipArt gallery: check all that are true.

- [ ] A. To open the gallery, go to ~~File ->Open~~ Insert clip art 2007
- [x] B. To use ClipArt, you would Insert a Picture from ClipArt
- [x] C. To find a horse you would type the word "Horse" in the Search For box
- [x] D. When you find the picture you want, you can drag and drop that picture into your document or just click on it

12. Working with ClipArt:
Microsoft Office has a ClipArt gallery: check all that are true.

- [ ] A. The ClipArt gallery cannot be used with spreadsheet or database, it is only for Microsoft Word.
- [ ] B. You cannot add new pictures to the ClipArt Gallery

13. Working with Graphics: check all that are true.

- [ ] A. Pictures float in a Frame
- [x] B. The picture frame has 8 handles, one on each corner and side
- [ ] C. To resize a ClipArt picture, you would use the four-headed arrow and drag it smaller
- [x] D. Pictures can be placed on top of each other in layers

14. Beginning Word: the Insert Menu
When you insert the date and time, you can make it automatically update.

- [x] A. True
- [ ] B. False

Microsoft Assessment Test

Beginning Word

15. Beginning Word: the Insert Menu
Insert a Picture from File lets you add digital photos to your document.

- [x] A. True
- [ ] B. False

16. Beginning Word: the Insert Menu 2003
You can also add photos by going to Insert ->File…

- [x] A. True
- [ ] B. False

17. Beginning Word: the File Menu
Check all that are true.

- [x] A. To open a new Word document, go to File on the menu bar then New
- [ ] B. To open a new Word document, go to File on the menu bar then Open
- [x] C. To make a copy of document with a different name, use File -> Save As

18. Beginning Word: the Edit Menu
Check all that are true.

- [ ] A. To remove a word or phrase and place it in another location go to Edit on the menu bar then Cut, Then Edit and Paste
- [ ] B. The menu command Edit -> Copy can be done with the keyboard command Control+C

19. Beginning Word: the Edit Menu
Which of the following commands are found under the Edit menu? Check all that are true.

- [ ] A. Undo
- [ ] B. Redo
- [ ] C. Cut
- [ ] D. Copy
- [ ] E. Paste
- [ ] F. Paste Special

or clipboard

20. Beginning Word: the Insert Menu
Which of the following commands are found under the Insert menu? Check all that are true.

- [x] A. Date/Time
- [ ] B. File
- [x] C. Picture
- [ ] D. Paste Special

Index Beginning Microsoft Word 2007 Exam 77-601 Guide